113 Topics in Current Chemistry

Fortschritte der Chemischen Forschung

Managing Editor: F. L. Boschke

Cyclophanes I

Editor: F. Vögtle

With Contributions by
V. Boekelheide, L. Rossa, I. Tabushi,
F. Vögtle, K. Yamamura

With 22 Figures and 19 Tables

 Springer-Verlag Berlin Heidelberg GmbH
1983

This series presents critical reviews of the present position and future trend
in modern chemical research. It is addressed to all research and industrial
chemists who wish to keep abreast of advances in their subject.

As a rule, contributions are specially commissioned. The editors and publishers
will, however, always be pleased to receive suggestions and supplementary
information. Papers are accepted for "Topics in Current Chemistry" in
English.

ISBN 978-3-662-15310-9   ISBN 978-3-540-40980-9 (eBook)

DOI 10.1007/978-3-540-40980-9

Library of Congress Cataloging in Publication Data. Main entry under title:
Cyclophanes.
(Topics in current chemistry = Fortschritte der chemischen Forschung; 113)
Bibliography: p. Includes index.
1. Cyclophanes--Addresses, essays, lectures.
I. Boekelheide, V. (Virgil), 1919 —. II. Series:
Topics in current chemistry; 113.
QD1.F58   vol. 113   [QD400]   540s [547'.5]   83-4650

© by Springer-Verlag Berlin Heidelberg 1983

Originally published by Springer-Verlag Berlin Heidelberg New York in 1983.

Softcover reprint of the hardcover 1st edition 1983

Introduction

The scientific importance of bridged aromatic compounds — PHANES — has continually increased during the last years, whereas previously they were considered as esoteric substances exclusive for academic interests.

The bridging of molecules using long chains or short clamps can be and has since been usefully applied in many branches of chemistry. This development was encouraged by several factors: New synthetic accessabilities for medium- and large rings were developed, which lead to phanes in up to excellent yields without using complicated techniques. The still widespread prejudice that larger rings generally imply difficult experiments and low yields should finally be eliminated. Besides multi-bridged, multi-layered, multi-stepped and helically wound phanes, compounds can be obtained in which the static and dynamic stereochemistry, steric interactions and electronic effects may be directionally adjusted and varied by bridge length, interior ring substituents and other parameters.

The spectroscopic methods allowed a rapid and detailed study of stereochemistry and electronic effects, among these bent and battered benzene rings, the face-to-face neighbourhood of aromatic nuclei, charge transfer effects and non-benzenoid π-systems. The availability of more efficient NMR- and mass spectrometers and X-ray analyses will admit more accurate interpretations for the future. Phanes may especially be exploited as model structures for weak intra- and intermolecular interactions.

It was only a small step from pyridinophanes to the complex-chemistry of crown- and cryptand-like ligands. Today it has expanded to ranges like capped cyclodextrins, bridged hem, concave host cavities for guest ions and -molecules as for water soluble enzyme models. In the latter field of receptors containing large cavities, pockets or niches, the development towards new branches of bioorganic and biomimetic chemistry has only just begun.

The present volume enters the field of PHANES with three research surveys, which are intended to fill at least some of the wide gaps since B. H. Smith published the "cyclophane bible" in 1964. The reviews mainly deal with the subgroup of the

CYCLOPHANES, which have been defined as those phanes containing *benzene* nuclei as ring members. Although the individual contributions of this (and following volumes) cannot include the entire phane chemistry, they give, however, a critical insight into actual theoretical and experimental developments, innovations and future trends.

After leaving its earlier isolation, the bridging of molecular skeletons has promoted most fields of chemistry with synthetic, stereochemical, spectroscopic and bioorganic relevancy. The literature quoted could moreover be useful for further interests in relation with phanes, crowns and other medio-/macrocyclic compounds as well as for their open chain analogues.

Bonn, October 1982   F. Vögtle

Table of Contents

Synthesis of Medio- and Macrocyclic Compounds by High Dilution Principle Techniques

Ludovica Rossa and Fritz Vögtle

Institut für Organische Chemie und Biochemie der Universität Bonn,
Gerhard-Domagk-Straße 1, D-5300 Bonn 1, FRG

Table of Contents

I Introduction

Within the last decades the chemistry of medio-[1] and macrocyclic ring systems, such as phanes, polyynes, annulenes, crown compounds, cryptands, ionophores, macrolides etc. has developed rapidly. Progress stems from an increased recognition of the theoretical [1-5] and practical importance [6-10] associated with these compounds and was aided also by a growing insight into general synthetic concepts and strategies for ring closure reactions. Among them, the *dilution principle* [4, 6a], the *template effect* [11], the *rigid group principle* [12], the *gauche effect* [11a], the *caesium effect* [13, 14] and other concepts [15] proved to be most profitable.

This contribution aims to give a detailed presentation of the different types of ring closure reactions used for the formation of medio- and macrocycles where the principle of high dilution plays a dominant role. Nevertheless, it should be kept in mind that in many cases discussed here, there is no strong decision hitherto between the effect arising from high dilution techniques and other concepts mentioned above.

Although the dilution principle approach plays the most important role in the synthesis of phanes, it is not restricted to them. In this survey in order to gain insight into the method and its detailed applications, we shall not limit the scope to phanes, but shall for the sake of comparison, include the other types of macro rings [16].

Cyclizations taking advantage of the dilution principle, usually start with open chained educt compounds bearing two or more functional groups, and, as a rule, only one of the possible oligomers, in most cases the monomeric cyclization product, is wanted as the main product. The formation of oligo- or polycondensation products usually is wished to be suppressed. The preference of the monomer formation is not simply based — as sometimes taken for granted — on the use of a large solvent volume and/or addition of highly diluted reagents. Contrarily, in "dilution principle reactions" not the total amount of the solvent volume is decisive, but instead the establishment of a stationary concentration of the educts in the reaction flask that is as low as necessary, to steer the cyclization reaction in such a way that ideally the same amount of starting material is flowing into the reaction flask per time unit as is reacted to yield the optimum of the target cyclization product [3, 4, 6a]. This flow equilibrium of educt influx and product outcome may be achieved with small volumes as well. We therefore use the expression "dilution principle reaction" ("DP-reaction") instead of simply speaking of "high dilution reaction" (DP = *Dilution Principle*).

The total amount of solvent may be diminished additionally by recycling the boiling solvents, which can be condensed to (pre-) dilute the starting components prior to their addition to the reaction flask. The simultaneous addition of components is facilitated by precision dropping funnels [6], pumps or syringes (s.b.).

Even today, the dilution principle is very often used empirically. Investigations into its theoretical and physico-chemical fundamentals have been undertaken much more rarely than synthetic and preparative studies. Nevertheless, there have been some

---

1 Regarding the term mediocyclic see: Thulin, B., Vögtle, F.: J. Chem. Res. (S) **1981**, 256 and references [33d, 34a, 65a, 232].

developments in theoretical directions [1-3], exceeding the former works of Ziegler [4] and Ruggli [5].

Recently Galli and Mandolini [1a] made the proposal of a classification of cyclization reactions, based on physico-chemical evidences resulting from e.g. lactonization reactions of ω-bromoalkanolates. Thereafter, the fundamental measure in cyclization reactions is the effective molarity EM. This molarity is defined as *the* reactant concentration, at which the intramolecular cyclization (k_{intra}) and the intermolecular formation of oligo-/polymeric products (k_{inter}) occur at the same rate ($k_{intra}/k_{inter} = 1$; k = rate constant). The course of the reaction depends on the initial concentration. If this concentration is small enough, the cyclization will dominate. In this case the educts can be put forward in the reaction mixture (batch-wise cyclization procedure).

As very dilute solutions are almost valueless in synthetic work, an "*influxion procedure*" is usually practical in preparative syntheses, where the reactant(s) is/are introduced slowly into the reaction medium over a longer period of time. The rate of feed v_f is the critical parameter now, which must be adjusted to make cyclizations dominant over polymerisation/polycondensation. This is achieved when $v_f < EM \cdot k_{intra}$. The rate of feed v_f substantially controls the duration of the process and it is a measure of its efficiency. This procedure corresponds to the Ruggli/Ziegler high dilution technique [4].

This review is mainly concerned with those reactions which yield preparative amounts of medium- and large ring systems and which therefore are generally carried out according to the influxion (DP-) procedure. In this study those reactions which were carried out according to the batch-wise technique and which are rare in the literature, are marked with the corresponding literature note [1a].

Though this theoretical progress is of general value, it is often of minor use in answering distinct preparative synthetic questions because in a cyclization reaction, the influence of several different, decisive reaction parameters must be taken into account (reactivity of the reactants, reaction temperature, reaction time, dilution ratio, solvent parameters, apparative factors). Such influences have seldom been studied in a systematic way by physico-chemical methods with respect to modern synthetic reactions [1, 2g, 3].

For reasons of limited printing space, full experimental description of characteristic reactions have, unfortunately, to be omitted here apart from some exceptions. Instead, only a few typical experimental parameters are sketched out informing solely on educts, products, type of reaction, solvent, dilution/predilution, time of addition, additional reaction time, reaction temperature, yield and references, data which should be useful in planning analogous and new DP-reactions as well especially for comparisons of substrate concentrations, solution ratios, yields etc.

One of these parameters, the type of reaction, may need some explanation [6a]: All DP-reactions are formally characterized according to the number of the educt components to be added (n-component DP-reactions, n = 1, 2, 3). If, for example, two educt solutions are dropped separately out of two dropping funnels or syringes into a certain solvent volume stirred in the reaction flask, we characterize this as a two-Component *DP*-reaction (2C-DP-reaction) [6a]. n in the nC-DP-reaction signifies only the number of the educt solutions dropped from each other independently into the reaction flask, and does not mean the total number of educts. Therefore, in the above example, even if an additional component has been dissolved in the reaction solvent in the reaction flask prior to the addition of the two components, or if it is mixed with one of the two educts in one of the dropping funnels, we nevertheless call this a 2C-DP-reaction, thus characterizing the technical procedure rather than chemical details. The latter may be noted in addition [6a].

The apparative handling of high dilution reactions is described elsewhere and can be omitted here; for up-to-date information see l.c. [6–10,23,24].

The following contribution attempts, therefore, to present a critical summary of the hitherto known facts and results on high dilution reactions; in addition it endeavours, as far as possible, to set up rules by comparative regard of reactions carried out according to the dilution principle, with respect to the type of reactions, dilutions and techniques applied. Such general comparisons and conclusions should be helpful in planning ring closure reactions. The philosophy of this report does not consist of a systematic presentation of all known DP reactions. It is, rather, a selection of characteristic cyclization reactions which allow a comparison with other synthetic methods and conditions in such a way that the generalizations and conclusions may help synthetic chemists to estimate the parameters, chances and yields of hitherto unknown cyclizations. Other publications, in which the synthesis of medio/macrocyclic compounds is described without remarks or details on the dilution conditions, are not considered in this text.

Hopefully, this — the first review on specific DP-reactions since Ziegler's [4] — may lead in future to a better calculation of this synthetic procedure, enabling scientists to carry it out in an optimal and standardized way and thereby saving time, needed otherwise for empirical testing.

II Nucleophilic Substitutions at Saturated C-Atoms

DP-reactions (DP = Dilution Principle) may be divided into sections according to the demand that they should proceed quickly, unambiguously and with high yield, at least with respect to the noncyclic model reaction using substrates exhibiting only one function each.

Such reactions are found in nucleophilic substitutions at saturated and unsaturated C-atoms. Apart from these, only few reaction types have been applied to the synthesis of medium/large membered ring systems in diluted solution. They will be discussed later, as the aforementioned are considered more important.

The nucleophilic substitution reactions can be further split according to the type of the attacking nucleophile (S, O, N and C-nucleophile). Most of the many DP-reactions can be systematically ordered according to this scheme. We start with C—S-bond forming DP-reactions, because many detailed studies have been carried out in this field and so comparisons between reactions, dilutions and yields can be drawn easier than with other reaction types.

II.1 Synthesis of Medio- and Macrocyclic Compounds by Formation of C—S-Bonds

Regarding ring closure reactions according to the dilution principle, including C—S-bond formation, only two sulphur delivering reagent types seem to have been used: *sulfide ions* generated from $Na_2S \cdot 9 H_2O$ or from thioacetamide, and on the other side, organic *thiols*. The latter are applied either as metal thiolates or as free thiols, which in situ form their metal salts in basic solutions. These methods have been used very often for the preparation of thiacycloalkanes, crown ether sulfides and thiaphanes.

The one starting component (substrate) not containing sulphur in most cases is a mono- or oligohalogeno compound. Often preferably bromo compounds have been used instead of the chloro compounds, because the C—Br-bonds are usually more reactive and bromide is a better nucleofuge than chloride. There seem, however, to exist exceptions in which chlorides lead to higher yields than bromides (cf. references at the appropriate place in the text below).

II.1.1 C—S-Bond Formation with $Na_2S \cdot 9 H_2O$

II.1.1.1 Thiacycloalkanes

The synthesis of thiacycloalkanes following the sodium sulfide method has been investigated only by a few groups [17-21] who prepared the eight- to fifteen-membered polymethylene sulfides principally according to the same two-component dilution principle reaction (2C-DP) [6a]:

Long-chain 1,ω-dibromo alkanes as a rule are cyclized with $Na_2S \cdot 9 H_2O$ [21]:

$$(CH_2)_n \begin{matrix} Br \\ Br \end{matrix} \quad + \quad Na_2S \cdot 9 H_2O \quad \xrightarrow{-2\,NaBr} \quad (CH_2)_n \quad S$$

1a–h	2	*3a–h*
n = 7–14		n = 7–14

Earlier works of Müller et al. [18] and Friedman and Allen [19, 20] show that this cyclization leading, e.g., to *3a–3h* can be carried out advantageously in a boiling solvent mixture (ethanol/water) of 2–5 l volume. The two starting components are dissolved in the same solvent mixture. Despite long addition times (15 hrs–2 days) and additional reaction times (2 hrs–11 days), the yields of the medium- and many-membered thiacycloalkanes *3a–3h* are rather low: 3–34 % [18-20].

Today, thiacycloalkanes can be obtained in substantially higher yields [21], if dipolar aprotic solvents as e.g. DMF, DMSO or HMPT are used as reaction medium. Especially HMPT has turned out to be favourable for the synthesis of *3a–3h*. As an example, we here give in detail a typical experimental procedure for the preparation of thiacyclopentadecane *3h*:

Into a stirred solution of 70 ml HMPT, 2.00 g (5.61 mmole) of 1,14-dibromotetradecane (*1h*) and equimolar amounts of $Na_2S \cdot 9 H_2O$ (*2*), dissolved in 15 ml of methanol each, are dropped by means of two dosing syringes. The addition is carried out over 2 hrs at 55 °C. At the same temperature, stirring is continued for 10 more minutes. After addition of 150 ml of H_2O, the reaction mixture is extracted with pentane for some hrs. The pentane extract is washed with H_2O and the organic layer dried over Na_2SO_4. After evaporation of the solvent, the raw product is purified by chromatographing on a column filled with silica gel. A 1:9 mixture of $CHCl_3$/petroleum ether is used for elution. The thiacycle *3h* is obtained in 43% yield (mp = 71 °C) [21].

Due to lack of space, in all further examples below we cannot give a detailed description of typical experimental procedures. Instead, a short insight into the essentials of the procedure with data in the following abbreviated manner will be given:

Experimental procedure for 1-thiacyclopentadecane (3h) [21]:
starting components: a) 1,14-dibromotetradecane (*1h*) (5.61 mmole) in 15 ml of methanol
 b) $Na_2S \cdot 9 H_2O$ (*2*) (5.61 mmole) in 15 ml of methanol
reaction type: 2C-DP-reaction [6a]
reaction medium: HMPT (70 ml)
reaction temperature: 55 °C
time of addition: 2 hrs
additional reaction time: 10 min
yield: 43% of *3h*

The other thiacycloalkanes *3a–3g* also are obtained in higher yields in HMPT: 78% for *3a*, 48% for *3b*, 28% for *3c*, 22% for *3d*, 25% for *3e*, 27% for *3f* and 55% for *3g* [21]. Not only the change of the solvent, but also the more refined technical experimental procedure seems to be responsible for the raised yields.

In the early 1950's, a dropping device had been developed specifically for ring closure reactions of thiacycloalkanes. It consisted of a lever made from a piece of yarn ("Wollfadenheber" [18b, 22]), with which the dropping rate could be regulated roughly under inert conditions. This proved to be necessary for Na_2S solutions, as the sulfur containing precipitate, which is formed during the often very long dropping times leads to irregularities or to a standstill of dropping. 10–20 years later, more satisfying precision dropping funnels [6b, 23] and/or syringe apparatus [8, 24] were used widely in high dilution reactions as they allow a comfortable and safe continous and synchronous inlet of one or more components.

It has long been general knowledge that the ring formation tendency is dependent upon the number of ring members [4] and that this is reflected in the yields. In addition to the entropy effects [1b, c, f] in the formation of medium rings, a yield minimum is often observed, for which transanular interactions [1d, f, 25] can be made responsible. This phenomenon known as "medium ring effect" [1f, 26, 27], is revealed clearly in the synthesis of oligothiacycloalkanes *3d–3f* containing 11–13 ring members: The corresponding yields are 22, 25 and 27% [21].

In this context, studies of Friedman and Allen [19, 20] are interesting in which negative influences of the medium ring effect are intended to be suppressed by the so-called "geminal dimethyl effect" [19, 20, 28].

The geminal dimethyl substituents at the C(5)- and C(6)-atoms of the 1,9-dibromo-5,5-dimethylnonane, 1,10-dibromo-5,5-dimethyldecane and 1,11-dibromo-6,6-dimethylundecane should bring the long-chain alkanes into conformations favourable to cyclization. Compared with the unsubstituted cycloalkanes, synthesized under the same conditions, the yields of the substituted rings show remarkable yield increases: from 5.7 [20], 13.0 and 3% [19] to 34 [20], 22 and 23% [19]. The geminal dimethyl effect hence has a decisive influence on the tendency of formation for the medium membered polymethylene sulfides.

In ring closure reactions leading to 1-thiacyclooctane (*3a*) and -nonane (*3b*) carried out by A. Müller and coworkers [18b], besides the monomeric products also the dimeric ones *4* and *5* have been isolated. The former were obtained in yields of 34 and 6.6%, the dimers of 3.9 and 6.7% [18b].

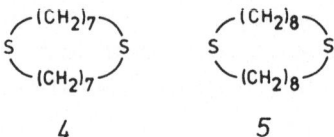

4   5

It should be pointed out here that by use of the caesium effect [13,14], a breakthrough in the field of the synthesis of mediocyclic sulfides has been achieved.

II.1.1.2 Crown Ether Sulfides

Of the large number of crown ether sulfides [29, 30, 32] only a few have been synthesized by use of $Na_2S \cdot 9\,H_2O$. Detailed experimental discriptions have been worked out for the crown ether sulfides 6–10 [30–32]:

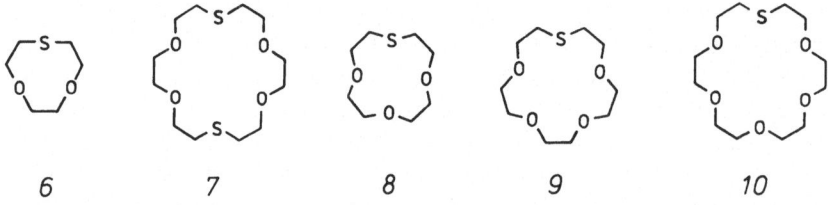

6   7   8   9   10

The cyclization reactions leading to these thiapolyethers exhibit some remarkable differences compared with the ring closure reactions yielding the thiacycloalkanes described in the previous Section II.1.1.1 [17–21]. In all syntheses of these medio/macrocyclic compounds open chained $1,\omega$-dichloro compounds are used as starting material, which is the only component. Dissolved in ethanol it is dropped into a certain volume of the same solvent [30, 32]. This proceeding has been called one component dilution principle reaction (1C-DP) [6a]. One mole $Na_2S \cdot 9\,H_2O$ is used for one mole of dichloro compound and predissolved in a volume of 600–1000 ml of ethanol together with a small amount of base (e.g. NaOH). The latter is intended to neutralize the weak acidic reaction of the sulfide anion [32b] to prevent a possible splitting of labile ether bonds.

Without attention to the dilution principle [1a], Dann et al. [32a] isolated in the ring closure reaction of 1,8-dichloro-3,6-dioxaoctane (11) in ethanol/water (1:1) with $Na_2S \cdot 9\,H_2O$ exclusively the dimeric product 1,10-dithia-4,7,13,16-tetraoxa-cyclooctadecane (7) in 2.6% yield. The monomeric 4,7-dioxa-1-thiacyclononane (6) has not been found in this cyclization reaction. Carrying out this reaction with consideration of the dilution principle, the isolation of 6 (5%) is possible. Also the dimeric product 7 was found in 12% yield [30].

$$\underset{11}{\begin{array}{c} \text{(structure with Cl)} \end{array}} \xrightarrow[\text{ethanol}]{Na_2S \cdot 9\,H_2O} \underset{7}{\begin{array}{c} \text{(crown ether sulfide)} \end{array}}$$

11   7

The three other products have been isolated after a reaction time of 7–14 hrs in the following yields: *8*: 14%, *9*: 29%, *10*: 36% [30, 32b].

A discussion and comparison of the yields of sulfur containing crown compounds will be given below, following the description of a further synthetic method, in Section II.1.4.1.

II.1.1.3 Thiaphanes

Phane systems of all ring types have been synthesized with the goal to study their stereochemistry and ring strain [1e, 33], the interaction between π-systems and the steric demand of intraanular substituents [34]. The importance of the synthesis of thiaphanes lies in the fact that they are easily available and in relatively high yields by ring closure reactions according to the dilution principle and that they are at the same time the cyclic precursors of the corresponding cyclophane hydrocarbons (carbaphanes). This can be achieved by extrusion of the sulfide sulfur which proceeds more easily than the extraction of N- and O-ring members. Several methods for sulfur extrusion are available: Stevens rearrangement [35], photo reaction of the sulfide in the presence of thiophilic phosphorous compounds [36] and the pyrolysis of the sulfone [37] which is easily obtained by oxidation of the sulfide.

Only a small number of the numerous thiaphane systems known today have been synthesized according to the Na_2S method. This is due to the fact that only intramolecular ring closure reactions to monosulfides or only symmetrical many-membered oligosulfides [38] are possible through intermolecular reactions. The attempted synthesis of unsymmetric thiaphanes starting with two different 1,ω-dihalogeno compounds would necessarily yield a mixture of products (cf. Sect. II.1.4.2).

Nevertheless, different structural types of thiaphanes have been prepared using Na_2S: hetero- and heterocyclic [33, 39] as well as single and morefold bridged [33, 39] ring compounds are known. The synthesis can be divided into IC-DP- and 2C-DP-reactions [6a]. In most cases, dibromo compounds have been used as one of the starting components, the other being $Na_2S \cdot 9 H_2O$, which is mostly dissolved in 50–95% ethanol/water; higher boiling apolar solvents have also been used.

2-Thia[3](2,2″)metaterphenylophane (*13*) [40a] was the first medium-membered monosulfide, which was submitted to the sulfone pyrolysis for ring contraction yielding the sulfur-free, strained[2](2,2″)metaterphenylophane (*14*) [40b].

| | 12 | 13 | 14 |

The ring closure reaction yielding the sulfide *13* has been carried out in ethanol as reaction medium and as the solvent for the raw viscous fluid dibromo starting material. 11% of the wanted monomeric thiaphane *13* were isolated after 5 hrs addition time and 12 hrs additional reaction time [40a].

In the framework of studies on steric interactions in the interior of ring systems, the following dithia[3.3]metacyclophanes *15a–e* [41)] have been synthesized according to the following abbreviated procedure:

	X	Y
15a	C	H
b	C	F
c	C	Cl
d	C	Br
e	N	–

50.0 mmole of the corresponding bis(bromomethyl) compound and 2,6-bis(bromomethyl)pyridine resp. in 250 ml of a solvent mixture of ethanol and benzene in a ratio of 4:1 were added dropwise simultaneously and synchronously with equimolar amounts of $Na_2S \cdot 9 H_2O$ in 250 ml of 90–95% aqueous ethanol to 1 liter of boiling ethanol over 4 hrs. After 12 hrs, the reaction was ended and the products *15a–e* were isolated in 8–10% yield [41)].

Some of these compounds were obtained later with higher yields by other research groups [42–45)]. The reason for the yield increases — 48% yield for *15a* [43)] and 19% yield for *15e* [42)] — is not easily apparent from the description. Parameters have been varied, but this should not have had an effect in the sense of a yield increase, as for example, in the synthesis of *15b* [44)]: Shorter dropping and no additional reaction time at the same temperature and using the same ethanolic medium resulted in a yield increase from 8–10% to 37%. The synthesis of *15e* is especially interesting, not only because the corresponding bis(*chloro*methyl) compound was used as a starting component, but also because, under otherwise equal conditions, 3% of the *tri*meric cyclophane [42)] with the exception of 19% of the *bis*-sulfide *15e* were obtained.

Compared to all other thiaphanes listed here, the 2,13-dithia[3.3](2,7)naphthalinophane (*16*) [46)] has been isolated in a high yield of 71%.

16

It is somewhat difficult to explain this result, because in a modification of the procedure used for the other metacyclophanes *15a–15e* [41–45)] described above, only benzene was taken as a solvent for the dibromo compound instead of ethanol and a somewhat shorter reaction time was used. It appears that the aromatic nuclei as rigid units exert a yield increasing effect (cf. rigid group principle [3, 4, 12)]). Further experimental comparisons are necessary to clarify remaining questions.

The synthesis of macrocyclic dioxo- [47)], tetraoxo- [48)], and hexaoxothiaphanes [49)] may be used for a comparison of the DP-parameters, e.g. 3,7-dioxo-5-thia[9](2,5)-furanophane (*17*), 1,5,12,16-tetraoxo-3,14-dithia[5.5]metacyclophane (*18*) and 1,5,12,-16,23,27-hexaoxo-3,14,25-trithia[5.5.5](1,3,5)cyclophane (*19*).

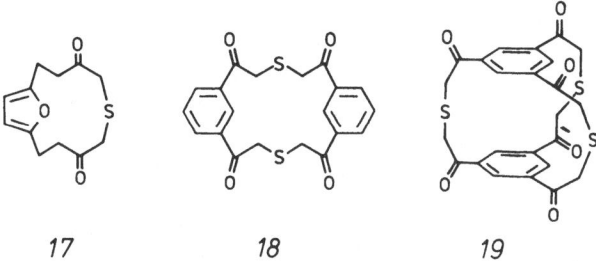

| 17 | 18 | 19 |

The bis(α-bromoacetyl) starting component for the synthesis of the furanophane *17* is soluble in THF and is dropped into ethanol simultaneously with ethanolic Na₂S solution over 8 hrs under boiling and continuous stirring. The intramolecular ring closure yields 43% of the monosulfide wanted [47]. In contrast to this, 1,3-bis(bromo-acetyl)benzene and 1,3,5-tris(bromoacetyl)benzene (the educts for *18* and *19*, resp.) are only soluble in unpolar solvents like benzene and are added dropwise to the more lipophilic, higher boiling tert-butanol over 8 hrs. This intermolecular ring closure reaction yields 5% of *18* [48] and only 0.2% of the cage molecule *19* [49]. One might draw the conclusion from this that the solvent for the educts can be chosen all the more unpolar, the more keto functions educts and products contain. But this has to be studied further using other examples.

As a consequence of choosing tert-butanol as a reaction medium, in which Na₂S and the bromoacetyl benzene mentioned above are sufficiently soluble, a higher reaction temperature results.

In going from *17* to *19*, a decrease in yield is observed, even if reaction times are extended (*17*: no additional reaction time; *18*, *19*: 8 hrs and 12 hrs additional reaction time resp.). This seems to be due to the number of bridges [33d] that have to be closed. As the ring closure of all three bridges is not proceeding at the same time, the possibility for undesirable side reactions grows.

The following thia- and dithiametacyclophanes [50, 51] have been synthesized according to the 1C-DP-procedure [6a]:

| 20 | 15a | 21 | 22 |

The preparations of the compounds *15a*, *20–22* are discussed separately, as they differ from the other [3.3]dithiacyclophanes in several reaction conditions. For the preparation of the 2-thia[3.2]metacyclophane (*20*), 3,3'-bis(bromomethyl)bibenzyl was added in the course of 18 hrs undiluted and in small portions to a mixture of methanol/water (30:1), in which an equimolar amount of Na₂S · 9 H₂O had been dissolved. The monosulfide *20* was isolated in 29.6% yield [50].

The phane *15a* has been prepared according to the 2C-DP-procedure [6a, 41, 43, 45]. The intermolecular ring closure reaction following the 1C-DP-procedure [6a] also starts with the same educt, 1,3-bis(bromomethyl)benzene, which in contrast to the above mentioned intermolecular 2C-DP-procedure [6a)] is dissolved in methanol. The latter solution is added to a solution of Na_2S (0.5-fold excess, relative to the educt) in the course of 72 hrs. The Na_2S is dissolved in a mixture of methanol/water (12:1). After the addition time of 72 hrs, the yield is 11.5% (*15a*) [50)]. Compared with the yield of 8–10% of *15a* [41)], no significant advantage of this IC-DP-reaction [6a)] is apparent.

The phane *15a* contains one more ring member and is therefore more flexible than the eleven-membered ring *20*. The phanes *21* and *22* have been synthesized starting with the corresponding methyl-substituted 1,3-bis(chloromethyl)benzenes, using an excess of Na_2S. Here, dioxane is the reaction solvent. The hexamethyl substituted compound *22* yields two isomers, exhibiting syn/anti-configuration due to the space demanding intraanular substitutents [50)]. Syn/anti isomerisation of *22* as described for the unsubstituted [2.2]metacyclophane itself [52)] is not possible because of steric hindrance.

The syntheses of other thiaphanes using the $Na_2S \cdot 9 H_2O$ method may be studied in the literature [53–57)].

Experimental procedure for 2,11-dithia[3.3]metacyclophane (15a) according to the $Na_2S \cdot 9 H_2O$ method [41)]:

starting components: a) 1,3-bis(bromomethyl)benzene (50.0 mmole) in 250 ml of ethanol/benzene (4:1)
                    b) $Na_2S \cdot 9 H_2O$ (*2*; 50.0 mmole) in 90–95% aqueous ethanol
reaction type: 2C-DP [6a)]
reaction medium: ethanol (1 l)
reaction temperature: boiling ethanol
time of addition: 4 hrs
additional reaction time: 12 hrs
yield: 8–10% of *15a*

II.1.2 C—S-Bond Formation using Thioacetamide

Thioacetamide has long been used as a reagent for the precipitation of metal sulfides. Thioacetamide, releasing sulfide ions, was utilized only a few years ago for cyclization reactions to sulfides in organic solvents [58)]. The thioacetamide method, therefore, is an alternative to the Na_2S method. Its advantage shall be demonstrated only for a few examples here; the syntheses of other thiaphanes using thioacetamide may be studied in the literature specified [59–62)].

The synthesis of the following thia- and dithiaphanes *15a* [41)], *13* [40a)] and *23* [57)] — the preparation of the first two [40a, 41)] using Na_2S has been described above (Sect. II.1.1.3) — is achieved starting with the corresponding bis(bromomethyl) compounds, by reaction with equivalent amounts of thioacetamide.

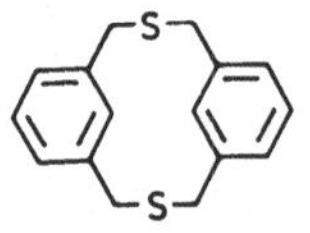

15a 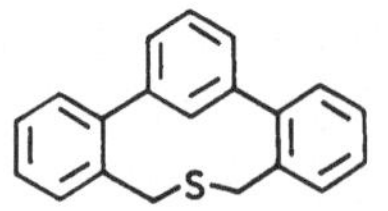   *13* 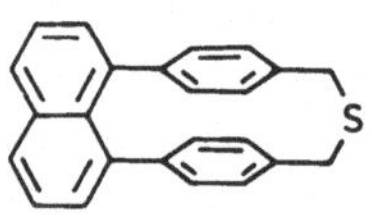   *23*

The weighing and dosing of the latter can be done without any difficulties, as thioacetamide is neither hygroscopic nor sensible to oxidation. The two components are dissolved in a mixture of polar and unpolar solvents. In the same volume a little more than the double molar amount of base is dissolved, which is nessecary for the formation of sulfide ions. The solvent has to be chosen in respect to the base used. For KOH and NaOH ethanol/water is suitable, for K-tert-butylate a mixture of anhydrous benzene and anhydrous ethanol can be taken. If K-tert-butylate is used as a base, cyclization reactions under anhydrous conditions are possible. Consequently, sulfides can be synthesized using thioacetamide, which are sensitive to hydrolysis. As Na_2S is insoluble in water-free ethanol, a small amount of water must be added in this case; therefore the Na_2S method is limited to products which are unsensitive to hydrolysis, and the addition of water can be disadvantageous because of the formation of heterogenous aceotropes. In this case dilution knees have to be used that are suited for heterogenous aceotropes. Usually, it is easier to have cyclizations done in a homogenous phase [58].

The further course of 2C-DP-reactions [6a] with thioacetamide as the source for thia sulfur differs scarcely from that of the thiaphane synthesis described above in Section II.1.1.3. The dropping times usually are somewhat longer in the thioacet-amide method, the reaction times — at the boiling temperature of the corresponding solvent — are usually somewhat shorter.

When comparing the yields of the thioacetamide and the Na_2S methods, the former seems attractive: in the cyclization to *15a* and *13* a yield increase from 8–10% [41] to 11% [58] and from 11% [40a] to 19% [58] was observed.

Experimental procedure for 2-thia[3](2,2″)metaterphenylophane (13) according to the thio-acetamide method [58]:

starting components: a) 2,2″-bis(bromomethyl)metaterphenyl (*12*; 2.00 mmole) and thioacetamide (2.00 mmole) in 150 ml of benzene/ethanol (2:1)
                        b) KOH (4.50 mmole) in 150 ml of ethanol/water (20:1)
reaction type: 2C-DP [6a]
reaction medium: benzene/ethanol 2:1 (1.2 l)
reaction temperature: boiling solvent
time of addition: 7 hrs
additional reaction time: 8 hrs
yield: 19% of *13*

The decrease of yields in the synthesis of compound *23* from 68% [57] to 48% [58] in switching over from the Na_2S to the thioacetamide procedure is as yet difficult to explain. Similarly, there is no adequate explanation for the increase in yields.

After many unsuccessful experiments [63–65], the synthesis of 2-thia(4,4″)orthoter-phenylophane (*25*) [14] with thioacetamide has been achieved — but only through application of the "caesium effect" [13, 14]:

               24   *25*

II.1.3 C—S-Bond Formation by Reaction of Metal Thiolates with Halogeno Compounds

Another method for the formation of C—S-bonds includes the nucleophilic substitution of halogeno compounds by metal thiolates. This procedure covers a broader spectrum of thia compounds than the C—S-bond formation using sulfide ions (sects. II.1.1, II.1.2), which is restricted to the synthesis of symmetric ring compounds. Numerous macrocycles containing sulfur have consequently been obtained using this method.

II.1.3.1 Thiacycloalkanes

The need to work under high dilution conditions [66, 67] is obvious when comparing the syntheses of some bis-sulfides. When cyclizing dibromo compounds with di-potassium-dithiolates without use of the dilution principle [1a, 66, 67], in addition to a high ratio of polycondensed products, the monomeric bis-sulfides 28 and the dimeric tetra-sulfides 29 are isolated only in low yields, never exceeding 4%, depending on the chain length of both reaction partners:

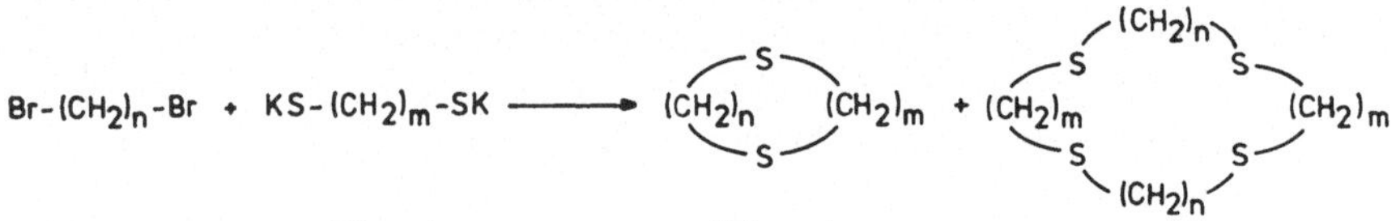

Br-(CH$_2$)$_n$-Br + KS-(CH$_2$)$_m$-SK

| 26 a-d | 27 a-d | 28 a-d | 29a-d |

	n			m
a	4		a	2
b	3		b	3
c	6		c	2
d	6		d	3

Depending on the number of CH$_2$ groups in the dibromo compound and in the dithiolate, n and m, the ratio of monomer and dimer amounts changes [66, 67]. With a ring number of 8 atoms, two different combinations are possible: n = 4 in 26a, m = 2 in 27a, and n = m = 3 in 26b and 27b. In the first case, no monomeric product 28a is obtained, and the dimeric 16-membered ring 29a is only formed in 0.2% yield. In the second case, n = m = 3, the monomer 28b is formed in 4% and the dimer 29b in 1% yield.

Of interest also are the reactions leading to 10- and 11-membered thia rings, as these lie in the so-called yield minimum [4, 6a, 67]. The 10-membered mediocycle 28c is formed as expected with an extremely low yield of only 0.06%, whereas the 11-membered bis-sulfide 28d could not be isolated at all. In the 11-membered ring series only the dimeric 22-membered tetra-sulfide 29d was obtained in merely 1.1% yield.

By adhering closely to the high dilution principle [4, 6a], the 10-, 14-, 15-, 18- and 22-membered bis-sulfides 32a–e have been prepared in a special cyclization apparatus [68]:

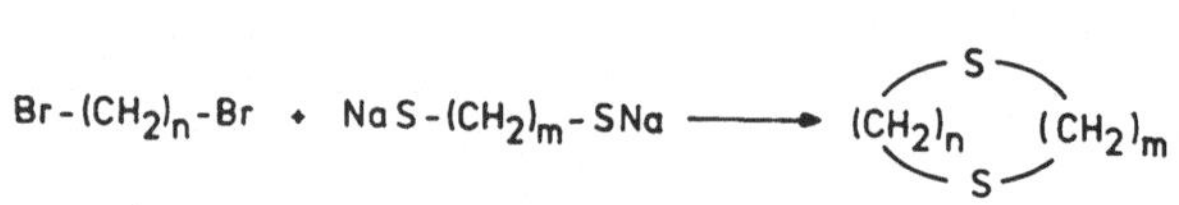

Br-(CH$_2$)$_n$-Br + NaS-(CH$_2$)$_m$-SNa

	n			m		
30a	2		31a	6		32a-e
b	6		b	6		
c	3		c	10		
d	6		d	10		
e	10		e	10		

No dimeric tetra-sulfides were obtained using this procedure. A comparison of the syntheses of the 10-membered thiacycloalkanes with and without application [1a] of the dilution principle seems to be appropriate: as mentioned above, the yield of the 10-membered thiacycle *28c* without application of the dilution principle, is only 0.06% [1a, 67]. If, in contrast, a solution of di-sodium-hexamethylene dithiolate (*31a*) in 500 ml dry ethanol and an equimolar amount of 1,2-dibromoethane (*30a*) in the same volume of water-free ethanol are added dropwise simultaneously and continuously to 800 ml of boiling anhydrous ethanol over 8 hrs, 34% *32a* are isolated after an additional reaction time of 1 hr. This drastic increase of ring formation, even in the range of the usual yield minimum, is strong evidence for the advantages derived from the dilution principle. The other thiacycloalkanes with larger ring widths are obtained in high yields: *32b*: 56%, *32c*: 41%, *32d*: 53%, *32e*: 69% yield [68].

Similar results [68] are encountered when the mixture of dithiol and dibromide and the equimolar amount of base (sodium ethanolate) are added from separate dropping funnels to the reaction medium (cf. Sect. II.1.4). This procedure is recommendable in all cases in which the thiolates are of low solubility.

In this field, the caesium method [13, 14] also seems to promote yields.

II.1.3.2 Thiaphanes

Numerous studies in the last years deal with the preparation of thiaphanes according to the thiolate method discussed here. A description of the success achieved in the last decade is not sufficiently detailed to draw comparisons here. Therefore, it is appropriate to also include some older investigations, which, for simplicity's sake we shall divide into the following three groups:

a) To start with, we shall give two synthetic procedures for thiaphanes which are of general character. A description of possible variations of the dilution principle conditions for the preparation of other thiaphane skeletons follows.

b) The following discussion includes those studies which focus on the dilution principle itself and its influence on the product formation.

c) Thirdly, we describe cyclization reactions according to the thiolate method which allow a direct comparison with other synthetic methods.

To a) above:

Many different types of thiaphane systems have been synthesized with the help of the thiolate method: morefold bridged [33a–d], hetero- and heterocyclic [39], symmetrical as well as asymmetrical ones (cf. Sects. II.1.1.3, II.1.2). Two syntheses which lead to asymmetrical products shall be chosen: the synthesis of the twofold bridged heterocyclic and heterocyclic systems *35* [69] and the preparation of the triply bridged trithia-phane *38* [49].

$$33 \qquad\qquad 34 \qquad\qquad 35$$

Experimental procedure for 2,9-dithia[2]metacyclo[2](2,6)pyridinophane (35) according to the dithiolate procedure [69]:

starting components: a) 2,6-bis(bromomethyl)pyridine (34) (50.0 mmole) in 250 ml of ethanol/benzene (4:1),
   b) 1,3-benzenedithiol (33) (50.0 mmole) and NaOH (100 mmole) in 250 ml of 95% ethanol

reaction type: 2C-DP [6a]
reaction medium: ethanol (1.5 l)
reaction temperature: boiling ethanol
time of addition: 3 hrs
additional reaction time: 12 hrs
yield: 29% of 35

Experimental procedure for 4,11,24-trioxo-2,13,22-trithia[4.4.4](1,3,5)cyclophane (38) according to the dithiolate procedure [49]:

starting components: a) 1,3,5-tris(bromoacetyl)benzene (36) (0.02 mole) in 250 ml of benzene,
   b) 1,3,5-benzene-tris(methanethiol) (37) (0.02 mole) and NaOH (0.06 mole) in 250 ml of ethanol/water

reaction type: 2C-DP [6a]
reaction medium: tert-butanol (2.5 l)
reaction temperature: boiling solvent
time of addition: 8 hrs
additional reaction time: 12 hrs
yield: 10% of 38

36   37   38

These two syntheses reflect the main features of a reaction according to the thiolate method that can generally be performed as 2C-DP-reactions [6a]. The dibromo compounds and the salts of thiols (dithiols, trithiols, . . .) are exclusively used as starting materials.

The dibromo compounds are, depending on their solubility, dissolved either in ethanol/benzene or in benzene alone or, additionally, in tert-butanol. As salts of thiols require a relatively polar solvent, in most cases water/ethanol mixtures are taken. In few exceptions, the thiolates are dissolved in a mixture of tert-butanol and ethanol. n-Fold molar amounts of sodium or potassium hydroxide are used as bases. Potassium hydroxide is preferred in combination with tert-butanol. The reaction medium, the volume of which is 1.5 l on an average, is adapted to the solvent or the solvent mixtures resp. All of these cyclizations have been carried out at the boiling temperature of the corresponding solvents, with the exception of a few reactions which proceed at room temperature [70, 71]. The influence of the reaction temperature on the product formation is not clear in each case. The dropping and reaction times differ greatly; they vary from 3.5 to 98 hrs. The yields differ likewise from 98 to 41%. The reactions can hardly be compared with each other without going into details.

The specific dilution principle conditions for the synthesis of further thiaphanes can be ascertained in the literature specified [34b, 36a, 44, 46, 49, 56, 63b, 70, 72, 73–80].

To b) above):

Only few studies deal with the dilution principle itself and the influence of the degree of dilution of the educts on the formation of different oligomeric skeletons [71, 74]. The reaction of the di-sodium salt of 1,3-benzenedithiol (33) with 1,3-bis(bromomethyl)benzene (39) indicates clearly the effect of the dilution principle [74]:

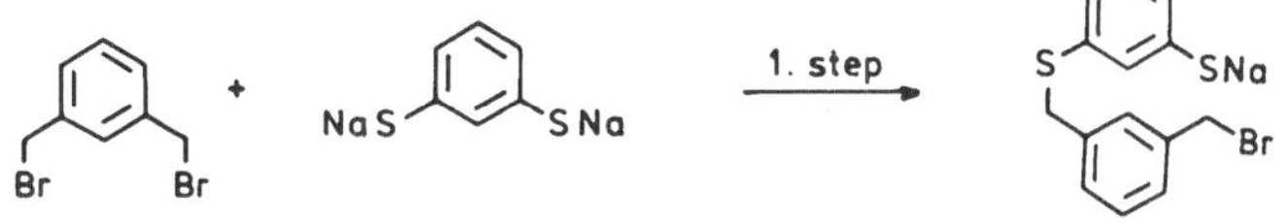

In this two-step reaction, the second step is decisive for the product type formed; in 40 only one of the two sulfide bridges is closed. If the further reaction proceeds intramolecularly, the monocycle 42 is formed exclusively. If the reaction course is intermolecular, an amount of oligo- and polycondensed reaction products in addition to the dimeric product 41 is obtained. The reaction 40 → 42 and 40 → 41 + 42 + polycondensed products can be directed into a particular direction by varying one or the other experimental conditions:

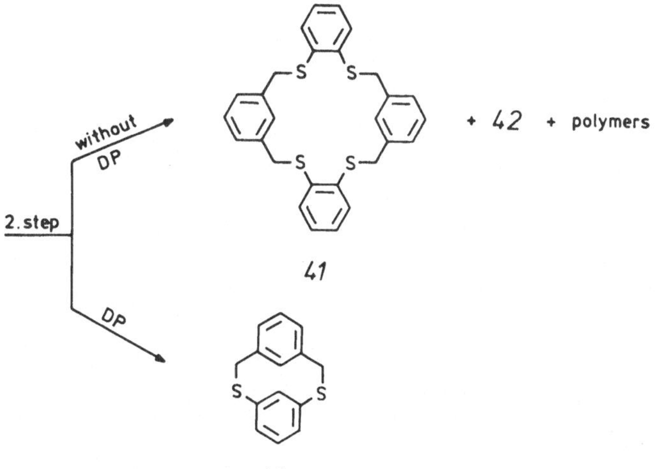

If the cyclization reaction 39 + 33 is carried out under strict dilution principle conditions according to the procedure described above, then the ring closure 40 → 42 proceeds intramolecularly [6a]: 1,10-dithia[2.2]metacyclophane (42) is obtained as a raw product in 98% yield! [74]. The rigid group principle [3, 11a, 12] is thought here to strongly favour the ring formation. If, under otherwise identical conditions, the reaction is carried out in more concentrated solution, e.g. 200 ml instead of 2.5 l tert-butanol as a reaction medium, then the yield of 42 decreases to 68% [74]. If the undiluted starting materials are dissolved in 2 l of ethanol and are heated together for several hours [1a], intra- and intermolecular ring closures compete with each other: 30% of 42 and 16% of 41 are isolated. The nevertheless relative high amounts of cyclic bis-sulfide 42 could well be due to the above mentioned effect of rigid groups [3, 11a, 12]. This principle seems to lose importance if an even smaller solvent volume (200 ml ethanol) is used: the yield of *42* (5%) decreases to less than that of *41* (15%).

Another study [71] concentrates on the influence of the product formation using different solvent types [2f, g, 3] and solvent volumes with respect to the dilution principle. The ratio of monomer to dimer formation is depicted in detail:

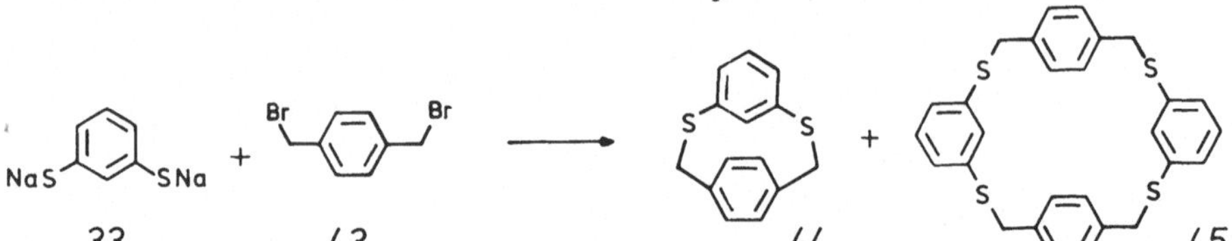

If tetrahydrofurane is chosen as a solvent for the dibromide *43* and ethanol for the di-sodium salt of *33*, and the 2C-DP-reaction [6a] is carried out as usual, the monomeric cycle *44* and the tetra-sulfide *45* are obtained in 6.9 and 3.4% yield resp. after a dropping time of 48 hrs into 1.5 l of boiling ethanol.

If a significantly smaller amount (only 200 ml) of the more lipophilic solvent THF is used as reaction medium, the yield of the dimeric product *45* increases to 5.7%. The formation of the dithia cycle *44* has not been detected under these circumstances, although the dropwise addition of the starting components to the reaction medium is significantly longer, namely 98 hrs.

The influence of the cyclization temperature (room temperature) on the product formation is not evident from the investigations [71]. Generally speaking, the more concentrated the cyclization solution is, the higher the amount of dimer, trimer, polymer will be. At the same time, the yields of monocycles decrease [1a].

To c) above:

9,18-Dimethyl-2,11-dithia[3.3]metacyclophane (*48*) can be synthesized in two different ways [75]: as it is a symmetrical thiaphane, it has been obtained according to the thiolate procedure or the Na$_2$S method (Sect. II.1.1.3).

Thiolate method:

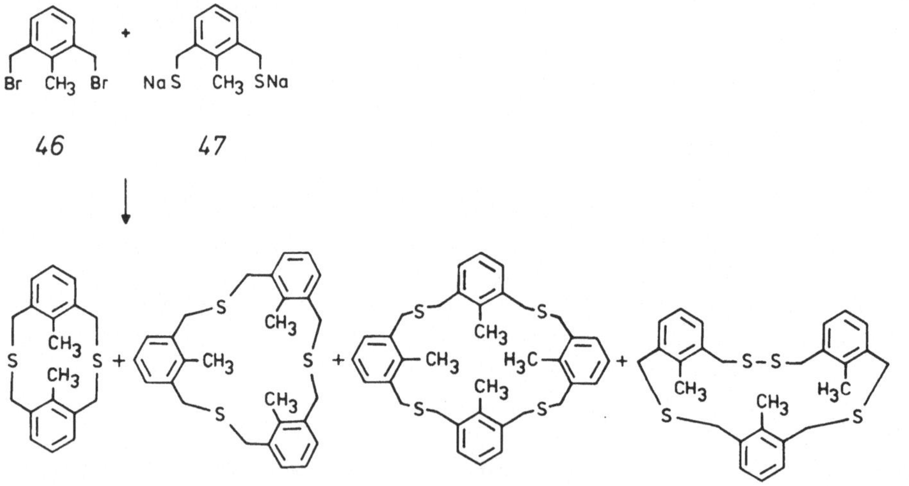

Na$_2$S method:

The solutions of *46* in a solvent mixture of ethanol/benzene 4:1 and *47* in ethanol/water are added dropwise as described above, to boiling ethanol in the course of 12 hrs. At the end of the reaction time (14 hrs), the raw product is worked up by chromatography. The residue was eluted using a 1:1 mixture of benzene/petroleum ether and gave four fractions. The first consisted of a mixture of the two syn/anti isomeric products *48a, b*. The separation of these products into the syn- and anti-isomers is possible at room temperature, because the intraanular methyl groups prevent a conformational equilibration of the isomers. The anti-isomer was formed in 15.7%, the syn-product in 1.4% yield. The second, third and fourth fraction of the original chromatography described above consisted of 1% *51*, 8% of the trimer *49* and 4% of the tetramer *50*. The formation of the trimeric bis-sulfide-disulfide *51* is not explained by the authors [75].

The same cyclization of 1-methyl-2,6-bis(bromomethyl)benzene (*46*), dissolved in benzene, with Na$_2$S · 9 H$_2$O in a 2C-DP-reaction [6a] (Sect. II.1.1.3) yielded only 6% anti- and 1.7% syn-isomer of *48* [75]. A comparison of the yields shows that the thiolate method is advantageous.

In the cyclization leading to 2,11-dithia[3.3]metacyclophane (*15a*), the advantage of the dithiolate method compared with the Na$_2$S method are even more pronounced: a 2C-DP-reaction using the dithiolate leads to 80% yield of *15a* [75], whereas under comparable conditions with Na$_2$S in a 2C-DP-reaction [6a] only 48% [75] could be obtained [41, 43] (cf. Sect. II.1.1.3). In this Na$_2$S cyclization, not only the desired *15a*, but also cyclic trisulfide (10%) and the tetrasulfide (2%) were formed [75].

II.1.3.3 Crown Ether Sulfides

The thiolate method has only rarely been applied to the synthesis of sulfur containing aliphatic crown compounds [81–84]. In most of them, all crown oxygen atoms are replaced by sulfur atoms [81]. The syntheses of these oligo sulfide rings in some essential DP-parameters differ from the procedures for the preparation of thiaphanes given in Sect. II.1.3.2 [49, 69, 71, 74, 75].

The 18-membered ring system 1,4,7,10,13,16-hexathiacyclooctadecane (*55*) was obtained by two groups in 1969 and 1974 [81, 82a]. The earlier study reveals that, according to method a), reaction of the educts *52* and *30a* in a 2C-DP-cyclization [6a] leads to 31% [82a] of the macrocycle *55*. The tetra-sulfide *56* and the 1,4-dithian (*57*) have not been detected in this study [82a].

Method a [81]:

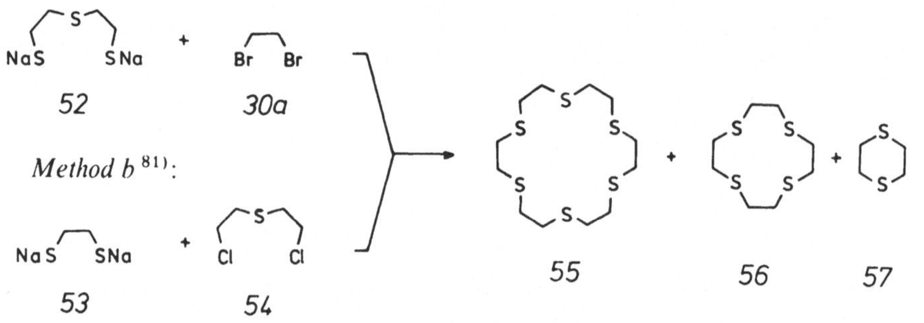

52   30a

Method b [81]:

53   54

55   56   57

For comparison's sake, methods a and b are sketched below:
Method a [81]:

1,2-dibromoethane (*30a*) and an equimolar amount of the di-sodium salt of 3-thiapentane-1,5-dithiol (*52*) are each dissolved in 500 ml of n-butanol. In the course of 2.5 hrs both solutions are simultaneously dropped into 2 l of the same solvent. During the first 10 hrs, the reaction mixture is cooled to 10 °C and then stirred for 36 hrs at room temperature. The yields are 8.1 % for *55*, 4.8 % for *56*, and 16 % for *57* [81].

Method b is described by the same authors [81]; it is carried out under identical conditions, but utilizes different starting components: the di-sodium salt of the 1,2-ethanedithiol (*53*) and the 1,5-dichloro-3-thiapentane (*54*). This reaction leads to *55–57* in 21.7 %, 1.2 % and 8.1 % yield resp. [81].

The discrepancies in the yields, especially with regard to the thia compound *55*, might possibly be due to varying characteristics of the starting components. As, according to method b, the ring closure to *55* is predominant, it has been concluded that chlorides in relatively apolar solvents favour the formation of the largest of all ring systems formed, but this postulate may not be generally applicable.

In the same study [81], further syntheses of macrocyclic sulfides are described, in which dichloro compounds are exclusively employed as starting components. It has, therefore, also been concluded that dibromo compounds disfavour the ratio of cyclic to polymeric products.

The formation of the mediocycle *56* and of the 6-membered ring *57* can not actually be anticipated from the reaction equation. The intermolecular ring closure of one molecule of each educt should lead to a nine-membered mediocycle containing three sulfur atoms, which, however, was not isolated in the cyclization reaction. The so-called "interchain-cyclization" mechanism has been proposed in explanation of the formation of *56* and *57* [81,84]:

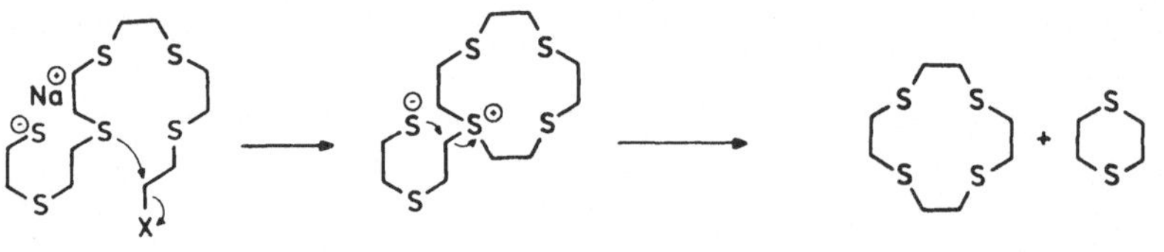

58   59   56   57

The synthesis of 1,4,7,10,13-pentathiacyclopentadecane *(61)* [81] is characteristic for the preparation of sulfur containing crown ethers according to the thiolate procedure:

starting components: a) 1,8-dichloro-3,6-dithiaoctane *(60)* (0.23 mole) in 400 ml of ether/n-butanol (1:1)

b) di-sodium salt of 3-thia-1,5-pentanedithiol *(52)* (0.23 mole) in 400 ml of n-butanol

reaction type: 2C-DP [6a]
reaction medium: n-butanol (2.2 l)
reaction temperature: below 60 °C
time of addition: 2.5 hrs
additional reaction time: 15 hrs
yield: 11 % of *61*

In contrast to the syntheses of thiaphanes (Sect. II.1.3.2), the educt often employed here is a dichloro compound. Also worth mentioning is the reaction temperature: whereas the ring closure reactions yielding thiaphanes [49, 69, 74, 75] take place in boiling solvents, essentially lower temperatures are used for the preparation of crown ether sulfides. The yields of the cyclic thia polyethers are moderate compared to thiaphanes and crown ethers themselves.

At this point, an interesting question arises: whether the caesium effect leads to an additional increase of yields here and if the favourable coordination (template) effect of sulfur with caesium is effective [13, 14].

A comparison between yield dependency on the ring number and the synthetic method will be drawn in the next Section II.1.4.1 in the course of a discussion on the fourth method for C—S-bond formation.

II.1.4 C—S-Bond Formation by Thiolates Produced in situ

A fourth method for the formation of C—S-bonds includes the nucleophilic substitution of a halogeno starting compound by a thiol in basic solution. The alkaline medium is necessary for the in situ formation of the thiolates. The attacking nucleophilic reagent is identical to that in the thiolate procedure. Whereas thiolates are soluble in relatively polar solvents only, this restriction does not hold for the neutral thiols, which, dissolved in apolar, more lipophilic solvents, can be added to the reaction medium. The addition of the components in the course of the thiol procedure can be carried out in four different ways: if the base is present in the customarily large solvent volume in advance, the cyclization can take place according to the 1C-DP- or the 2C-DP-method [6a]. It is advantageous to add both educt components, the halogeno compound and the thiol, using only one single dropping funnel, in order to circumvent the difficulties involved in simultaneous dosing. This is possible only if the two educts display similar solubility characteristics and if the starting components do not react in the (precision) dropping funnel itself. If the base is not present in the reaction flask in advance, but dissolved in a suitable solvent and added dropwise to the reaction medium, two alternatives arise: one can proceed according to either the 2C-DP- or to the 3C-DP-method [6a]. In both cases, the addition must be made simultaneously.

The thiol method is suited for the synthesis of asymmetrical cyclic products as is also the thiolate method (Sect. II.1.3). Only the C—S-bond formation using sulfide ions is limited to the preparation of symmetrical ring compounds (Sects. II.1.1 II.1.2).

The thiol method is the one most generally applicable for C—S-bond formations. Basically, it can replace all three aforementioned methods of C—S-bond formation (Sects. II.1.1, II.1.2, II.1.3). The dilution parameters for thiacycloalkanes according to the thiol method are the same as those for thiaphanes. Only the syntheses of phane systems are discussed here.

II.1.4.1 Crown Ether Sulfides

The synthesis of macrocyclic polyether sulfides has often been described [30, 32b, 66, 67, 81–84a]. The experimentation is still mostly concerned with forming cation complexes and other typical properties of sulfide sulfur bearing crown compounds and is rather less detailed with regard to the cyclization reaction and to aspects of the dilution principle.

One of the procedures [32a] for the synthesis of 1,10-dithia-4,7,13,16-tetraoxacyclo-octadecane (7) according to the thiol method is discussed here in more detail because it has been carried out both with and without application [1a] of the dilution principle:

Equimolar amounts of 3,6-dioxa-1,8-octandithiol (62), 1,8-dichloro-3,6-dioxa-octane (11) and sodium carbonate are dissolved in 1 l of 50% ethanol/water and refluxed for 64 hrs: 34% of the macrocycle 7 desired are isolated [1a], the analytical data of which are identical to those of the product obtained in the cyclization reaction of 11 with Na₂S (cf. Sect. II.1.1.2) in 2.6% yield [32a].

If, in contrast to this procedure, the mixed components 62 and 11 are added dropwise from a single dropping funnel over 1 hr into a stirred boiling solution of 4 l of 50% ethanol/water containing an equimolar amount of Na_2CO_3, the 18-membered bis-sulfide ring 7 is obtained as expected in the higher yield of 58.5%. This yield increase again demonstrates the effectiveness of the dilution principle. The cyclization according to the thiol method is not only a substitute for the Na_2S method, but, considering the higher yield, is also the preferred method for the synthesis of 7.

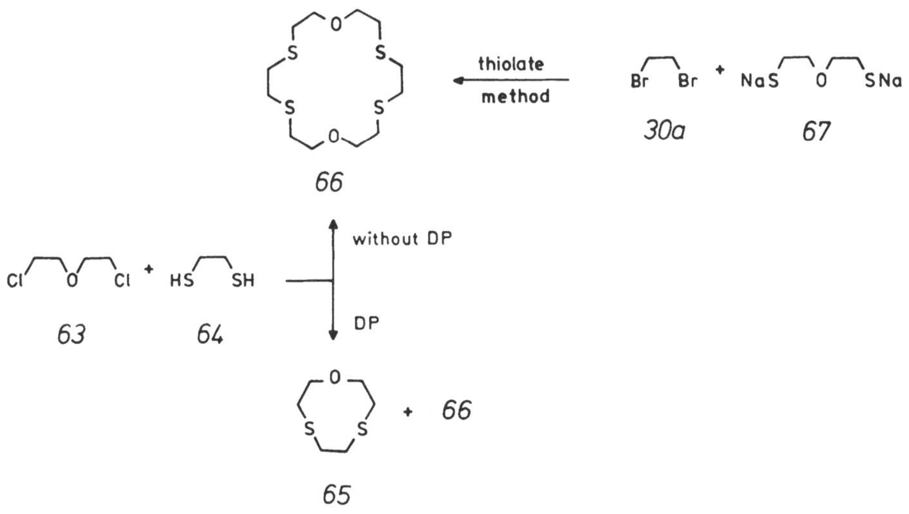

The experimental implementation of these reactions is decisive in the formation of the different ring systems 65 and 66. In 1934, Meadow and Reid [85] attempted to synthesize the 9-membered dithia compound 65 by a reaction of 63 and 64. This synthesis failed because the dilution principle was disregarded [1a]: they were only able to isolate the dimeric product 66. Under dilution conditions according to the thiolate method [82a] (Sect. II.1.3.3) and thus proceeding from starting with other educts [1,2-dibromoethane (30a) and the di-sodium salt of 3-oxa-1,5-pentanedithiol (67)], the synthesis of 65 again was not successful: solely the macrocyclic crown tetrasulfide 66 was formed in a 7% yield. As late as 1972, Bradshaw et al. [30] succeeded in preparing 7-oxa-1,4-dithiacyclononane (65) and 7,16-dioxa-1,4,10,13-tetrathiacyclooctadecane (66) according to the thiol method: the educts 63 and 64 were added separately to one liter of ethanol containing NaOH under strict observance of all the conditions of a 2C-DP-reaction [6a]. Consequently, 65 was isolated in a 6% and 66 in a 4% yield.

The isolated products 66 [30, 85] obtained in accordance to the thiol method with and without application of the dilution principle [1a, 4], have identical melting points of 125–126 °C, whereas the product obtained according to the thiolate method [82a] exhibits a much lower melting point of only 50 °C. The assumed structure for the macrocyclic compound 66 was confirmed by all three authors [30, 82a, 85] by analytical data. An explanation for the melting point difference is still outstanding.

At this stage of the discussion, a general comparison of the yields of all cyclic polyether sulfides [32b, 66] with the cycloaliphatic compounds to be discussed later is appropriate, especially with respect to the dependence of the ring member number [86]. The 8-11-membered aliphatic ring systems synthesized through dinitrile condensation (Thorpe-Ziegler condensation, cf. Sect. III.3.2) and through Dieckmann condensation (cf. Sect. III.3.1) have been obtained in very low yield, or they are not formed at all (yield minimum [4]). With increasing ring member number (12 and higher), the yields of the cyclic hydrocarbons increase. Similar tendencies are observed in the series of the corresponding ether sulfides, but the yields of all ring widths lie somewhat higher, especially in the series of the mediocycles.

The cyclization of 1,9-dicyanononane [87] producing the 10-membered hydrocarbon only leads to 0.5% of cyclic product. A remarkable yield increase from 0.5 to 5% was achieved [32b, 87] when a methylene group was exchanged for an oxygen atom in the center of the aliphatic chain. Consequently, hetero atoms in an aliphatic chain may decrease the flexibility of the latter and are advantageous to the conformation of the ring closure. Thus it is not surprising that the synthesis of 4,7-dioxa-thiacyclononane (6) [30] by cyclization of 1,8-dichloro-3,6-dioxaoctane (11) with Na_2S (Sect. II.1.1.2) led to a 5–6% yield.

Following the introduction of rigid groups [3, 11a, 12], e.g. aromatic nuclei, the conformational flexibility of the hydrocarbon chain containing hetero atoms is even more restricted and thus the yields are even better. Sulfide ring systems of this type have been synthesized in the last several years [88].

The interrelationship between the flexibility, ring width and the melting points of ring systems has been investigated in some detail by Dale [89].

Finally, a typical experimental procedure shall be given for the preparation of oxathia ring compounds according to the thiol method: the synthesis of 4,10,13-trioxa-1,7-dithiacyclopentadecane (69) [30]:

starting components: a) 3-oxa-1,5-pentanedithiol (68) (0.5 mole) in 300 ml of ethanol
                     b) 1,8-dichloro-3,6-dioxaoctane (11) (0.5 mole) in 300 ml of ethanol
reaction type: 2C-DP [6a]
reaction medium: ethanol (1 l), excess of NaOH
reaction temperature: 6 hrs (time of addition) room temperature, 4 hrs (additional reaction time)
                     boiling solvent
time of addition: 6 hrs
additional reaction time: 4 hrs
yield: 27% of 69

II.1.4.2 Thiaphanes

The addition of the starting components may be carried out in four different ways, as has previously been mentioned in the introduction of the thiol procedure (Sect. II.1.4). The classification of high dilution reactions into 1C-, 2C- and 3C-DP-reactions [6a] is used here also for the syntheses according to the thiol method, whereby the 2C-DP-additions [6a] can again be subdivided into two groups. As an introduction to each group we give short notes on detailed experimental procedures for the preparation of thiaphanes, after which we describe possible variations of the DP-parameters. As far as data published allow comparative contemplations regarding other addition procedures and synthetic methods they will be presented.

In contrast to the synthesis of thiapolyethers, which usually begin with chlorides, preparation of thiaphanes more often requires the use of bromo compounds as starting components. Only in a few cases are chlorides [90, 91] used, especially for the synthesis of multilayered [33d] and multi stepped [33d] cyclophane hydrocarbons. These are constructed by repetition of the following steps: a) cyclization to the thiaphane and b) desulfurization. The first ring closure yielding the thiaphane is usually accomplished by a reaction of the dichloro compound with dithiols. After the sulfur atoms have been extruded, the second cyclization starts with dibromides. No explanation for this change from chloro- to bromo compounds is given in the corresponding studies [90, 91].

1C-DP-reaction [6a]:

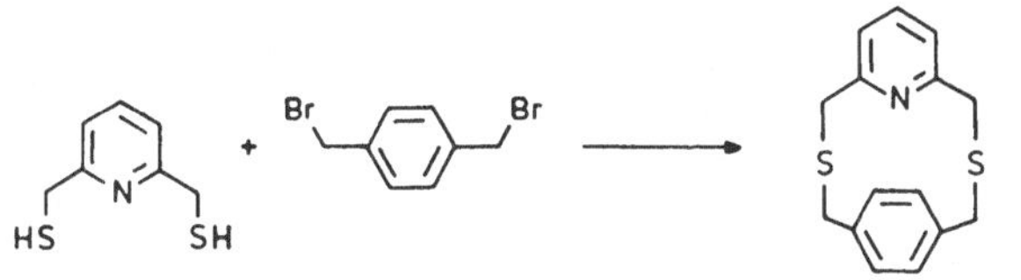

70   43   71

Experimental procedure for 2,11-dithia[3.3](2,6)pyridinoparacyclophane (71) [92]:
starting components: 2,6-pyridino-bis(methanethiol) (70) (26.0 mmole) and
                1,4-bis(bromomethyl)benzene (43) (26.0 mmole) in 500 ml of benzene
reaction type: 1C-DP [6a]
reaction medium: ethanol (2 l) containing 50.0 mmole of NaOH
reaction temperature: room temperature
time of addition: 12 hrs
additional reaction time: 10 hrs
yield: 48% of 71

The advantage of a 1C-DP-reaction [6a] has been noted above: the simultaneous dosing from two or three dropping funnels is obviated. This advantage (of the 1C-addition) is put to use in the majority of phane syntheses done according to the thiol method [93a – l]. Usually, benzene is chosen as a solvent for the starting dibromo compound and dithiol. In some cases the more polar dioxane [93h] is applied. NaOH or KOH in water/ethanol serve as bases; they are often employed in some excess.

Only rarely have reactions according to the thiol procedure been carried out at room temperature [90, 92, 93j, k] — as in the above mentioned example. Boiling solvents are characteristic not only of 1C- but also of the 2C- and 3C-cyclizations [6a]. Large differences are observed in addition and reaction times as well as in the yields. The addition and reaction times range from a minimum of 8 hrs [93j] to a maximum of 5.5 days [91]. Yield discrepancies range from 100% [90] to 37% [93h].

Remarkably, 71 can also be synthesized according to the Na₂S method [92]:

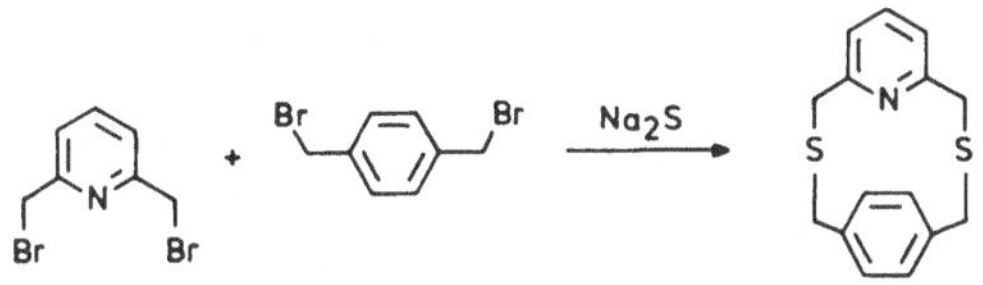

34   43   71

starting components: a) 2,6-bis(bromomethyl)pyridine (34) (32.3 mmole) and 1,4-bis(bromomethyl)-
                benzene (43) (32.3 mmole) in 900 ml of benzene
               b) Na₂S · 9 H₂O (200.0 mmole) in 900 ml of ethanol
reaction type: 2C-DP [6a]
reaction solvent: ethanol (1.5 l) and benzene (500 ml)
reaction temperature: no instructions
time of addition: no detailed remarks
yield: 36% of 71

Though *71* is an asymmetrically composed phane, the yield is relatively high. The formation of symmetrical cycles has not been observed. No explanation for this remarkable fact is given by the authors [92] (Cf. Sect. II.1.1.3).

2C-DP-reaction [6a]:

a) In the first subgroup of these reactions, the educts are added separately from two dropping funnels. The solutions of the starting components are dropped into a solvent volume containing a third basic component. An example for this type of reaction is the following cyclization [94]:

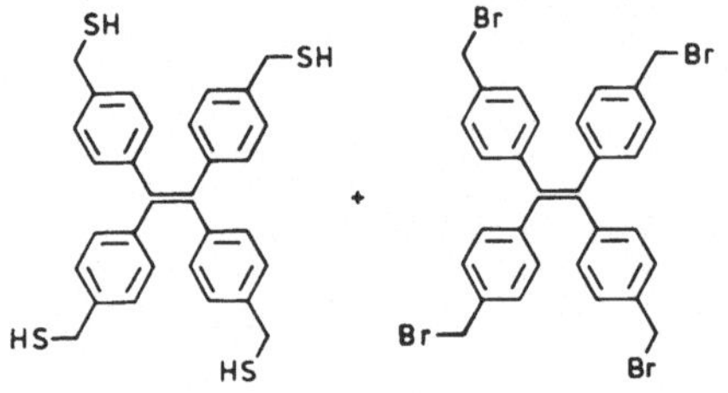

72   73

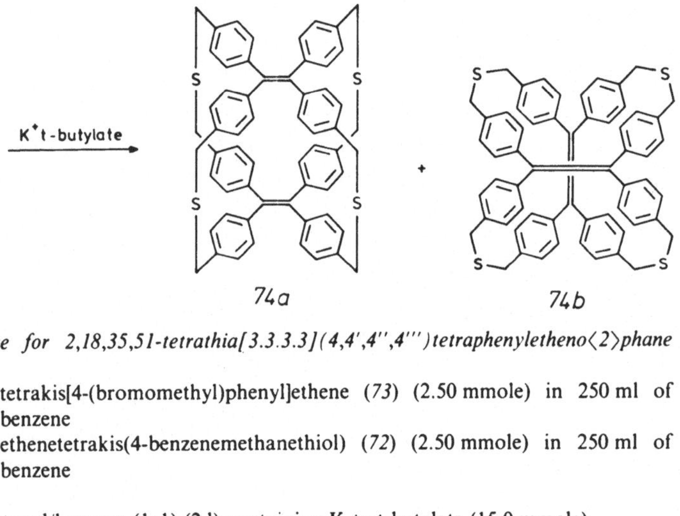

K⁺t-butylate →

74a   74b

Experimental procedure for 2,18,35,51-tetrathia[3.3.3.3](4,4',4'',4''')tetraphenyletheno⟨2⟩phane (74a,b) [94]:

starting components: a) tetrakis[4-(bromomethyl)phenyl]ethene *(73)* (2.50 mmole) in 250 ml of benzene
                   b) ethenetetrakis(4-benzenemethanethiol) *(72)* (2.50 mmole) in 250 ml of benzene

reaction type: 2C-DP [6a]
reaction medium: tert-butanol/benzene (1:1) (2 l) containing K-tert-butylate (15.0 mmole)
reaction temperature: boiling solvent
time of addition: 7 hrs
additional reaction time: 1–2 hrs
yield: 4%, a mixture of the isomeric cycles *74a,b*

This experimental procedure [94] has not generally been applied; in other syntheses according to the thiol method (2C-DP-reaction [6a]), other solvents, such as toluene [98], ethanol, THF [96,98], DMF [96], and their mixtures [96,98] and other bases, such as KOH [95,96b,97] NaOH and K_2CO_3 [96a], have been used. The above cyclization also is comparable to a 3C-DP-reaction [6a,94]. In this procedure, KOH dissolved in ethanol is added as a third component simultaneously with the other two benzene solutions of *72* and *73* to a benzene/ethanol (2:1)-mixture; this variation did not lead to a yield increase (4% mixture of isomers *74*). The low yield may be due not only to the number of bridges to be closed, but also to the low solubility of an intermediate cyclization product.

b) A second subgroup of the 2C-DP-reactions [6a] covers only a few syntheses of monocyclic thiaphanes, the formation of which is usually achieved by simultaneous addition of two solutions: one dropping funnel containing the dibromo compound and the dithiol, the other dropping funnel containing the base in a suitable solvent. An example of this type of addition is the preparation of 2,5-dithia[6.1]metabenzeno-phane-13-one (*76*) [99]:

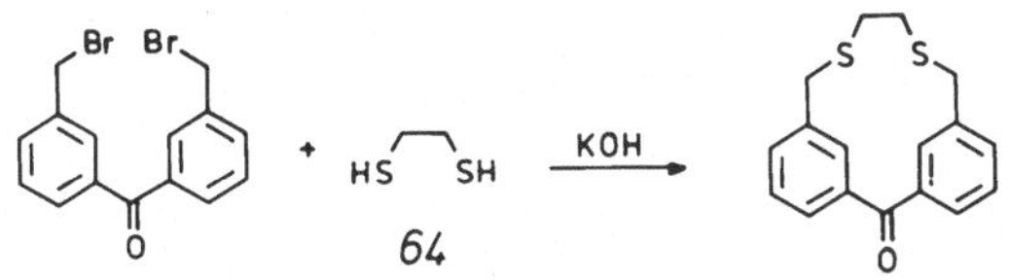

starting components: a) 1,2-ethanedithiol (*64*) (10.0 mmole) and 3,3'-bis(bromomethyl)benzophe-
     none (*75*) (10.0 mmole) in 250 ml of benzene
     b) KOH (15.0 mmole) in 250 ml of ethanol
reaction type: 2C-DP [6a]
reaction solvent: ethanol (2 l), benzene (400 ml)
reaction temperature: boiling solvent
time of addition: 3–4 hrs
additional reaction time: 1 hr
yield: 62% of *76*

Some cyclic intermediates for helical hydrocarbons have been synthesized according to this method [100].

3C-DP-reaction [6a]:

Cyclizations leading to thiaphanes by separate addition of three different components are almost unknown. Nevertheless, such reactions have led to satisfactory yields, often by use of three precision dropping funnels [23] and of specially constructed 3C-DP-apparatus [23].

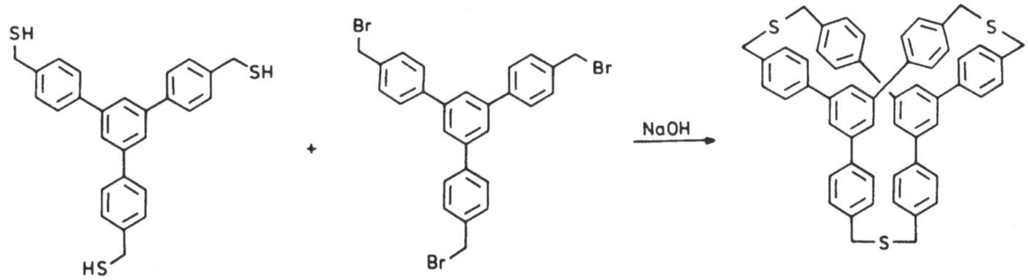

Experimental procedure for the synthesis of 2,23,50-trithia[3.3.3](4,4',4'')-1,3,5-triphenylbenzeno-phane (79) [101]:
starting components: a) 1,3,5-benzene-tris(4-benzenemethanethiol) (*77*) (8.54 mmole) in 250 ml of
     DMF
     b) 1,3,5-tris[4-(bromomethyl)phenyl]benzene (*78*) (8.54 mmole) in 250 ml of
     benzene
     c) NaOH (25.8 mmole) in 250 ml of ethanol

reaction type: 3C-DP [6a)]
reaction medium: benzene/ethanol (1:1) (1.5 l)
reaction temperature: boiling solvent
time of addition: 8 hrs
additional reaction time: 1 h
yield: 31% of *79*

The cyclization reaction using Na$_2$S (cf. Sect. II.1.1.3) as the source of the sulfur component, dissolved in ethanol/water (1.5:1), and the tribromo compound *78*, dissolved in benzene as a second starting component, only yielded 8.6% of the trithiaphane *79* desired [101)]. This shows the superiority of the thiol method over the C—S-bond formation using Na$_2$S. The difficulties of the simultaneous addition of three components can be overcome by use of specially designed 3C-DP-apparatus [23)] The problems of dosing Na$_2$S solutions have been mentioned in Section II.1.1.1.

Further syntheses of other thiaphane systems according to the 3C-DP-method [6a)] can be found in the original literature [63b, 102, 103)].

II.2 Synthesis of Medio- and Macrocyclic Compounds by C—O-Bond Formation

The formation of C—O-bonds using alcoholates in a nucleophilic substitution step at the saturated C-atom proceeds slower than the C—S-bond formation owing to the higher nucleophilic power of the thiolate compound compared with the alcoholate anion [104)]. Such slower proceeding cyclizations are — as has been mentioned before (cf. introduction of Sect. II) — less suitable for reactions according to the dilution method. Nevertheless, most C—O-ring condensations proceed with unexpected good yields and often without applying the dilution principle [1a)]. This may be due to the coordinative effect of the oxygen atoms [11a–i, 13, 14, 15i, j)] in the acyclic educts which favours the formation of medio- and macrocyclic products. Metal ions, present either as inorganic bases for the formation of the alcoholates or which must be added to the reaction medium, coordinate with the oxygen atoms and restrict the conformational flexibility of the oligooxy-chain. This template effect [11a–h)] forces or facilitates a favourable orientation for the ring closure step.

Equally effective in the formation of oxa cycles are rigid groups [3, 11a, 12)] such as double and triple bonds as well as aromatic nuclei in the open chained starting components. In many syntheses of polyethers both effects (the template effect [11a–i, 13–15i, j)] and the rigid group principle [3, 11a, 12)]) contribute to the yield increase of the cyclic products.

Medio- and macrocyclic compounds which possess more than one oxygen ether functions, can be classified as crown ethers or oxaphanes. An exact allocation may be somewhat difficult in certain cases, especially if there are several condensed aromatic or cyclohexano rings [29)]. Though the first polyethers were synthesized as early as 1937 by Lüttringhaus and Ziegler [106, 107)] according to the high dilution method, crown ether chemistry received its impetus as late as 1967 through the work of Pedersen [105)]. The properties of crown-type compounds — the selective and stable complex formation with inorganic and organic cations [108–111)] and anions [112, 113)] as well as with uncharged organic molecules [114)], the ability to transport cations through membranes and the applicability as phase transfer reagents and catalysts [115] and kinetic studies [1e] — are some of the reasons for their intense study in the last years. They lead to a manyfold structure variation of ether oxygen containing medium- and large ring compounds. Their syntheses have been reviewed several times [29a, e, 116–120]. They can be divided in two large groups: Preparation without [1a] and with application of the dilution principle [4]. The latter principally also are obtained according to the Williamson ether synthesis [121].

II.2.1 Polyethers according to Williamson's Ether Synthesis

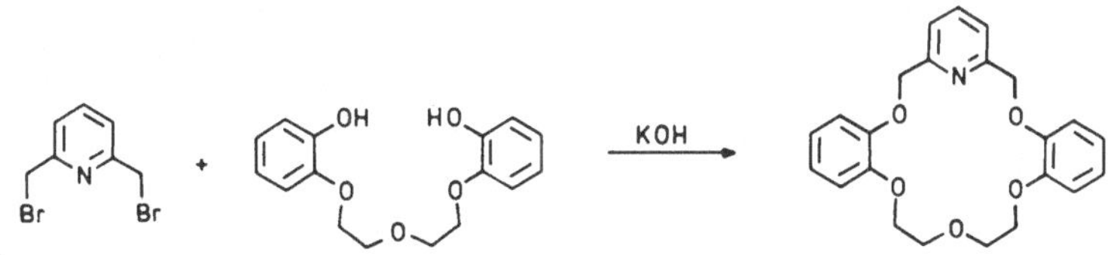

34   *80*   *81*

Experimental procedure for 18⟨O₅(2,6)pyridino-(1,2)benzeno.2₂.(1,2)benzeno.1.1-coronand-6⟩ *(81)* [122] *or 1,4,7,14,23-pentaoxa[7]orthobenzeno(2,6)pyridino[2]orthobenzenophane (81)* [39b, 123]:

starting components: a) 2,6-bis(bromomethyl)pyridine (*34*) (20.0 mmole) in 250 ml of benzene
                     b) 1,5-bis(2-hydroxyphenoxy)-3-oxypentane (*80*) (20.0 mmole) in 250 ml of DMF
                     c) KOH (40.0 mmole) in 250 ml of ethanol/water (50:1)
reaction type: 3C-DP [6a]
reaction medium: n-butanol (1 l)
reaction temperature: boiling solvent
time of addition: 8–10 hrs
additional reaction time: 2 hrs
yield: 30% of *81* [123]

Macrocyclic polyethers may also be prepared according to the 1C-, 2C- or 3C-DP-procedure [6a] (cf. thiol method for the synthesis of thiaphanes, Sect. II.1.4.2). Apart from bromo- and chloro compounds tosylates often are used as starting components, the tosyl group being accepted as a good leaving group. NaOH, KOH, NaH and potassium tert-butylate are used as bases, n-butanol, tert-butanol, DMSO and THF and glymes are suitable as solvents. The cyclizations are carried out at room temperature or in boiling solvents.

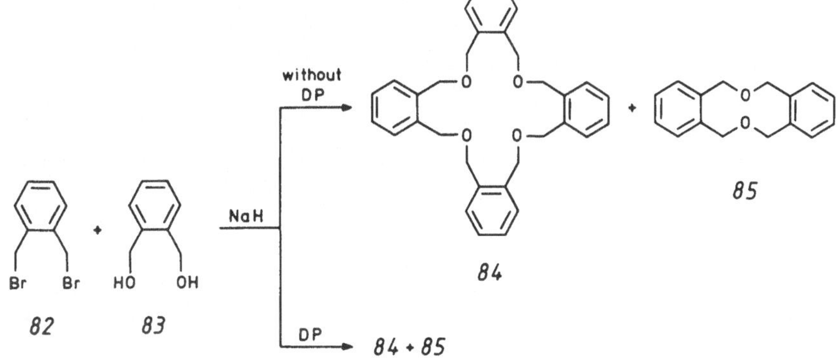

The yield proportions of the cyclic oligo ethers *84* and *85* can be steered in the above reactions [124]: if 1,2-bis(bromomethyl)benzene (*82*) is reacted with the di-sodium salt of 1,2-benzene-bis-methanol (*83*), the dimer of *85*, 2,11,20,29-tetraoxa-[3.3.3.3]orthocyclophane (*84*) is isolated as a main product in a 40% yield [124], whereas the desired product *85*, the 2,11-dioxa[3.3]orthocyclophane is formed only in a 15% yield [124].

If the reaction conditions are changed and the dilution principle is strictly observed, the yield ratios are shifted in favour of cycle *85*. If equimolar amounts of *82* and *83* are dropped simultaneously into a suspension of NaH in inert solvent, *85* and *84* are formed in 55 and 7% yield resp. [124].

A marked ring size effect for both of the high dilution cyclization steps involved in the synthesis of multi-loop crown ethers [125a – c] has been demonstrated to depend on the chain length of the oligoethylene glycol ditosylates used.

II.3 Synthesis of Medio- and Macrocyclic Compounds by C—N-Bond Formation

II.3.1 C—N-Bond Formation by Reaction of Amines and Sulfone Amides with Halogeno Compounds

Nucleophilic substitution reactions at saturated C-atoms with amines under C—N-bond formation have seldomly been carried out under dilution principle conditions. More important are reactions yielding nitrogen containing medio- and macrocyclic systems using acid amides as intermediate products that are reduced to the free amines (cf. Sect. III.2).

Substitution reactions of alkyl halides using primary amines easily lead to tertiary amines, because the secondary amines as a rule are more nucleophilic than the original acyclic primary amines. In intramolecular reactions, high reaction speeds are often desirable [126]. If, on the other hand, cyclization should lead only to the secondary amine, as is often the aim in intermolecular ring closure reactions, then the reactivity of the primary amino function must be reduced. This can be achieved by the formation of the sulfone amides. Most often para-toluene sulfone amides have been used, more rarely the mesyl amides.

II.3.1.1 Azacycloalkanes

The first syntheses of azacycloalkanes, especially in the medium membered ring area, were unsatisfactory, as in some cases, they were carried out without applying the dilution method [1a]. Also, the experimental techniques of high dilution reactions were not highly developed [127 – 131]. Although azacycloalkanes, especially those of medium ring size, are today more readily available by reduction of the corresponding cyclic acid amides, the preparation according to the nucleophilic substitution at saturated C-atoms nevertheless seems of some interest with regard to the specific formation of uniform dimeric cyclic compounds, the diazacycloalkanes.

Under this aspect, we here discuss a reaction [131], which, strictly taken, is not implemented or carried out according to the dilution principle [1a] (cf. introduction, Sect. I), but which, nevertheless, when in a highly diluted solvent [5, 132a, b] and with apportioned addition of the educts, leads to diazacycloalkanes. Variation of some of the important reaction parameters influences the product formation markedly. It was expected that the formation of the 18-membered diamine *87* was preferred over the ring closure to the 9-membered cycle.

First, 1,8-dibromooctane (*1b*) and 1,8-octanediamine (*86*) × 2 HCl after addition of double molar amounts of NaOH or Na_2CO_3 were cyclized in 5 l of 50% ethanol containing NaOH or Na_2CO_3. The dimeric base *87* was isolated in 11.7% yield, whereas in more concentrated solution (using 2.9 l instead of 5 l 50% ethanol) and otherwise following an analogous procedure, *87* was obtained in only 9% yield [131].

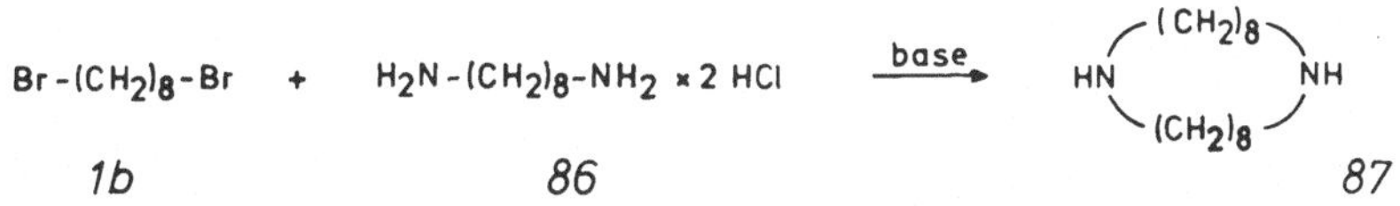

$$Br-(CH_2)_8-Br \quad + \quad H_2N-(CH_2)_8-NH_2 \times 2\ HCl \xrightarrow{base}$$

$$\begin{array}{c} \overset{(CH_2)_8}{\diagup\quad\diagdown} \\ HN \qquad NH \\ \diagdown\ (CH_2)_8\ \diagup \end{array}$$

$$\quad\quad 1b \quad\quad\quad\quad\quad 86 \quad\quad\quad\quad\quad\quad\quad\quad\quad\quad 87$$

As an intermolecular ring closure always requires at least two steps, in this reaction the acyclic bromo compound *88* is formed first. In more concentrated solution, a part of the bromide *88* is substituted by hydroxy anions, if NaOH is employed as a base, to give the primary alcohol *89*, in which the abstraction of H_2O to yield the desired cycle *87* is no longer possible [131].

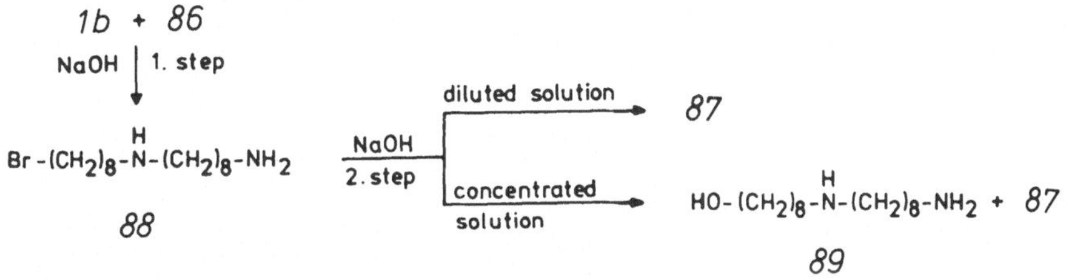

$$1b \ + \ 86$$

NaOH $\downarrow$ 1. step

$$Br-(CH_2)_8-\overset{H}{N}-(CH_2)_8-NH_2 \xrightarrow[\text{2. step}]{NaOH}$$

88

diluted solution → *87*

concentrated solution → $HO-(CH_2)_8-\overset{H}{N}-(CH_2)_8-NH_2 \ + \ 87$

89

Sodium carbonate is better suited as a base; under otherwise identical experimental conditions in diluted as well as in concentrated solution, the 18-membered macrocycle *87* could be obtained in 17% yield [131].

The tosyl derivative of *86* allows even better results with regard to *91* starting with 1,8-dibromooctane (*1b*) and 1,8-octanediamine-ditosylate (*90*) in pentanol using a threefold amount of water-free K_2CO_3 [131, 133].

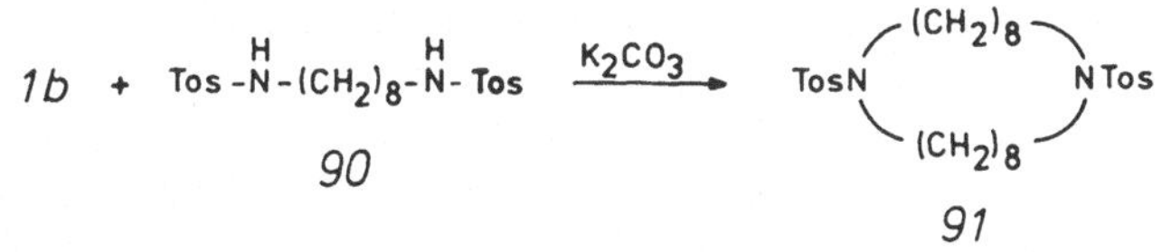

$$1b \ + \ Tos-\overset{H}{N}-(CH_2)_8-\overset{H}{N}-Tos \xrightarrow{K_2CO_3} \begin{array}{c} \overset{(CH_2)_8}{\diagup\quad\diagdown} \\ TosN \qquad N\,Tos \\ \diagdown\ (CH_2)_8\ \diagup \end{array}$$

$$\quad\quad\quad\quad\quad\quad 90 \quad\quad\quad\quad\quad\quad\quad\quad\quad\quad\quad\quad 91$$

The experimental procedure for this ring closure reaction is similar to a cyclization according to the 2C-DP-method [6a]:

starting components: *1b* (6.25 mmole) in 56 ml of pentanol, *90* (6.25 mmole) undissolved (in 28 portions)
reaction solvent: pentanol (125 ml), containing threefold amount of potassium carbonate [131]
reaction time: 9 hrs
yield after detosylation of *91*: 30% of *87*

Additional experiments in larger scale only reached yields of 20% for 87 [1a, 131].

The 8-, 16- and the 10-, 20-membered N-containing cycloalkanes have been synthesized under dilution conditions [134]. The corresponding bromo alkyl sulfone amides [129] had been chosen as starting material and added dropwise to a suspension of K_2CO_3 in n-pentanol (1C-DP-reaction [6a]). N-para-toluene sulfonyl-azacyclo-octane (93) is obtained in 60% yield. On a smaller experimental scale and using longer dropping times exceeding 26 hrs even higher yields were obtained [134].

The dimeric ring compound 94 has been detected, but in such small amount that no yield percentage has been noted.

The N,N'-ditosyl-1,9-diazacyclohexadecane (94) nevertheless has been synthesized starting from other components and using another solvent: the educts 1,7-dibromo-heptane (1a) and p-toluene sulfone amide (95), dissolved in methylethylketone, were dropped into 370 ml of the same solvent over a period of 72 hrs, containing K_2CO_3 in excess: yield 17% of 94 [134].

$$Br-(CH_2)_7-NHTos \xrightarrow[\text{n - pentanol}]{K_2CO_3} \quad (CH_2)_7 \quad N-Tos \quad + \quad Tos-N \begin{matrix}(CH_2)_7\\(CH_2)_7\end{matrix} N-Tos$$

$$92 \qquad\qquad\qquad 93 \qquad\qquad\qquad 94$$

$\uparrow$ K_2CO_3 methylethylketone

$$Br-(CH_2)_7-Br \quad + \quad H_3C-\langle\text{C}_6H_4\rangle-\overset{O}{\underset{O}{S}}-NH_2$$

$$1a \qquad\qquad\qquad 95$$

$$Br(CH_2)_9 NHTos \xrightarrow[\text{n - pentanol}]{K_2CO_3} \quad (CH_2)_9 \quad N Tos \quad + \quad TosN \begin{matrix}(CH_2)_9\\(CH_2)_9\end{matrix} N Tos$$

$$96 \qquad\qquad\qquad 97 \qquad\qquad\qquad 98$$

The cyclization of 96 yielding N-tosyl azacyclodecane (97) and its dimeric analogue 98 was led through analogously to the 1C-DP-reaction [6a], starting with 96 [134], but there is found another ratio: The azacyclodecane 97 could only be isolated in 3%, whereas the dimer 98 was obtained in 11% yield [134]. Benzyl alcohol as a solvent only leads to polymeric products [134]. The yield decrease from the 8- to the 10-membered ring (93: 60%, 97: 3%) is in accord with the above mentioned lower tendency of formation of mediocycles (yield minimum [4]) that favours the formation of the dimeric macrocycles.

The first macrocyclic compound containing four nitrogen atoms and 20 ring members has been synthesized by Stetter and Roos [135] according to the dilution principle. The ring closure reaction started with an acyclic dibromo compound, already containing two C—N bonds; the second component was a N,N'-ditosyloligomethylene diamine. The tetraaza ring *101* has been formed by coupling only two C—N bonds:

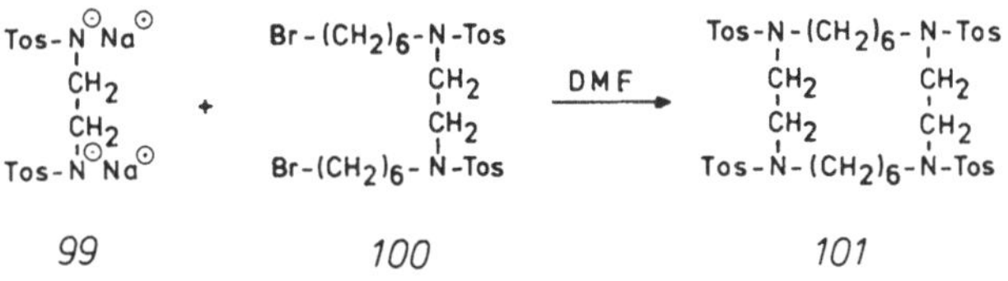

| 99 | 100 | 101 |

Experimental procedure for N,N',N'',N'''-tetratosyl-1,4,11,14-tetraazacycloeicosane (101) [135]:

starting components: di-sodium salt of N,N'-ditosyl ethylene diamine (*99*) [136] (0.02 mole) in 400 ml of MeOH/DMF (1:1) and N,N'-ditosyl-N,N'-bis(6-bromohexyl)ethylene di-amine (*100*) (0.02 mole) in 400 ml of DMF

reaction type: 1C-VP [6a]

reaction medium: DMF (900 ml)

reaction temperature: boiling DMF

time of addition: 30 hrs

additional reaction time: 1 hr

yield: 39% of *101*

Stetters concept [135] for the preparation of tetraaza macrocycles [138] has been applied successfully e.g. in the structure determination [137] of an optical active alkaloid, pithocolobine.

II.3.1.2 Azaphanes

Only a few C—N-bonds leading to azaphanes are formed by nucleophilic substitutions at the saturated C-atom. Like azacycloalkanes, azaphanes are mostly prepared by nucleophilic substitutions with amines at *unsaturated* C-atoms and subsequent reduction of the carbonyl function. Aside from a few exceptions, which will be discussed below, the synthesis of all N-containing phane systems by nucleophilic substitution at the saturated C-atom starts with N-tosylamine derivatives and follows essentially the same principle. We therefore give a general experimental instruction for the synthesis of the simple diazaphane *103* [139].

Experimental procedure for N,N'-ditosyl-N,N'-decamethylene-p-benzenediamine (103) [139]:

starting components: N,N'-ditosyl-p-benzenediamine (*102*) (20.0 mmole) and 1,10-dibromodecane (*1d*) (20.0 mmole) dissolved in 350 ml of DMF

reaction type: 1C-DP [6a]

reaction medium: 350 ml of DMF containing K$_2$CO$_3$ (0.30 mole)

reaction temperature: 130–135 °C

time of addition: 21 hrs

additional reaction time: 2.5 hrs

yield: 45% of *103*

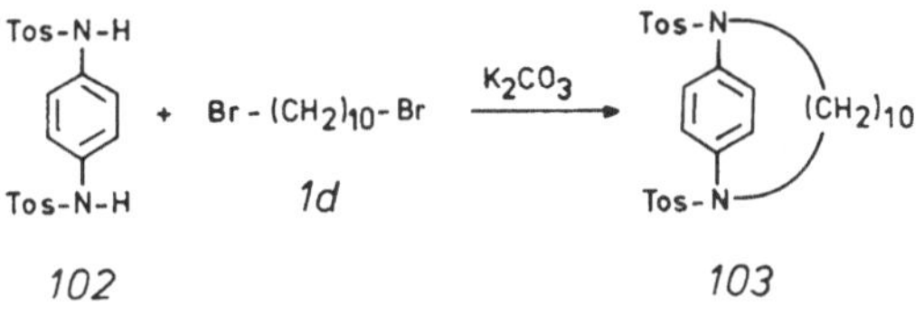

| 102 | 1d | 103 |

Typical for the preparation of azaphanes are the following conditions:

a) all C—N-ring closure reactions have been carried out according to the 1C-DP-method [6a]; b) the bromo compound is one of the ring forming components; c) K_2CO_3 as a base is used in large excess; d) DMF is mostly employed as solvent [139, 140a, b]; also used as solvents are pentanols [142, 147-150] and DMSO [141].

The following shows other methods of preparations: bis-(N,N'-ditosyl-N,N'-pentamethylene-para-benzenediamine) (106) [135] was prepared in analogy to the synthesis of the 20-membered aliphatic macrocycle 101, which was described in Section II.3.1.1 [135]. Not the free sulfone amide but the di-sodium salt of 104, formed in advance of the cyclization, is reacted with the 1,5-dibromo pentane derivative 105. Consequently, the addition of base, which is indispensable for the formation of the alkali salts of sulfone amides in situ, is unnecessary for this type of ring closure reaction.

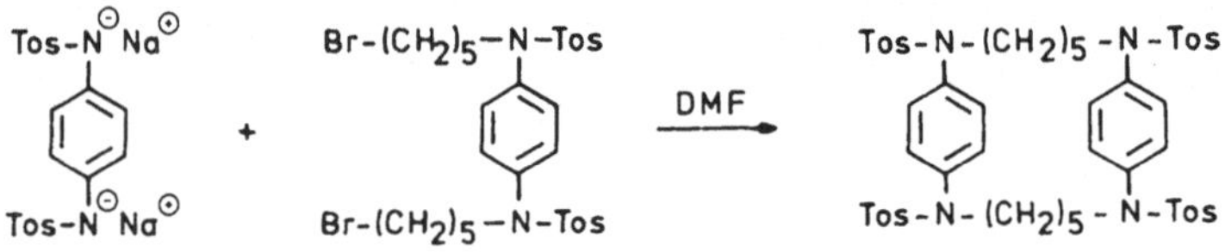

104   105   106

The 22-membered tetraaza macrocycle 106 could be isolated in 40% yield [135]. The advantages of this procedure were explained in Section II.3.1.1 [135].

Remarkable in this context is that experiments to carry out the cyclization to the benzidine tetraazaphane 109a–c according to the same method were wrecked on the low solubility of the di-sodium salt of 108 [143]. Consequently, the dibromo compound 107, necessary for the cyclization, could not be obtained.

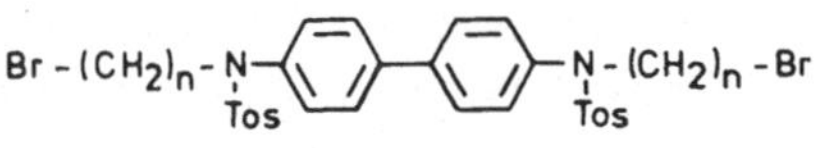

107

To circumvent this problem, the free sulfone amides 108 were cyclized with the dibromo alkanes 30a, 26a, b according to the above mentioned procedure in DMF using K_2CO_3 as a base.

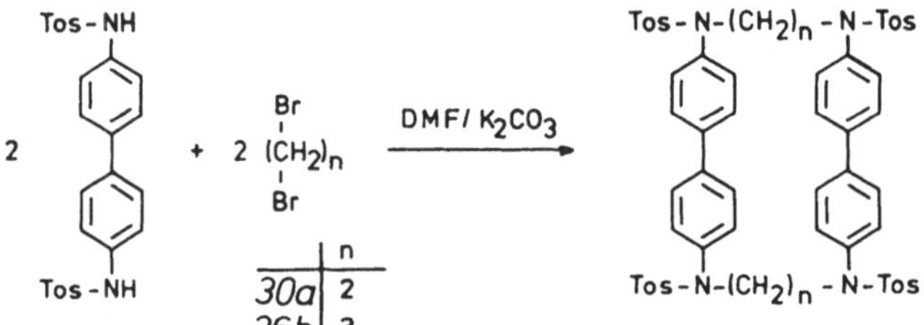

	n
30a	2
26b	3
26a	4

108

109	n
a	2
b	3
c	4

The macrocyclic systems *109a–c* thus occurred in 15 (*109a*), 20 (*109b*) and 24% yield (*109c*)[143]. These relatively satisfactory yields — taking into account the cyclization of four reaction partners — may be based on the rigid biphenyl units[143] (rigid group principle[3,11a,12]).

Without use of the dilution principle[1a] also the 7-11-membered mediocycles containing aromatic nuclei were unexpectedly available with quite good results[144,145], which may be explained similarly by the rigid group principle[3,11a,12].

Under the aspect of host compounds suitable for inclusion of guest molecules, the N,N',N'',N'''-tetratosyl-1,6,20,25-tetraaza[6.1.6.1]paracyclophane (*110*)[146] was recently synthesized in 25% yield in a similar way as *109c*[143].

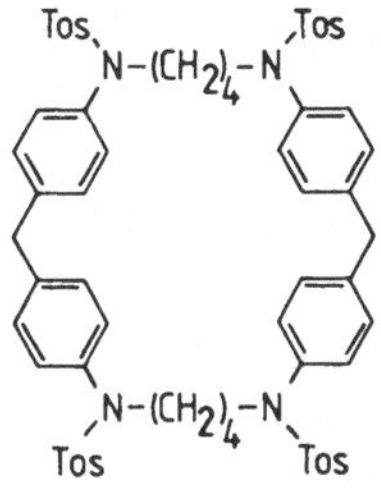

110

The fourfold protonated tosyl-free tetraaza compounds derived from *109c*[143] as well as those of *110*[146] enclose solvent molecules; the decisive proof, an X-ray analysis, has been carried out with the detosylated tetraazaphane *110* as a tetra-protonated host and durene as the guest molecule, located in the centre of the host molecules cavity[146].

The aimed synthesis of e.g. [2]-, [3]- and higher catenanes[126,147–151] has been achieved in different ways. Here only one of the different synthetic methods for [3]catenanes shall be viewed; it includes an intra- and an intermolecular C—N-ring closure step[150,151b]. It was assumed that for the formation of the di-ansa compound *112* the two intramolecular alkylating ring closure reactions at the aromatic NH$_2$ function should attack once above and once below the plane of the aromatic ring. This steric demand if fullfilled in a spiro compound that is obtained by acetal formation at a catechol unit[151b]:

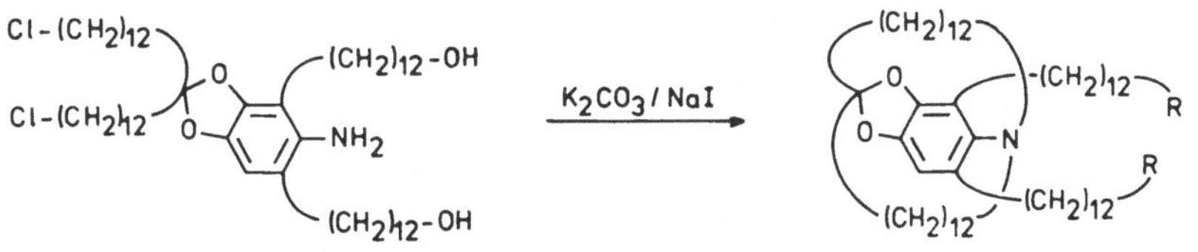

111

	R
112	OH
113	Br
114	NH Tos

The experimental procedure of this cyclization differs in several parameters from the general method mentioned above: The formation of a tertiary amine is desired here (cf. Sect. II.3.1). The chloro compounds are transformed into the acetyl protected derivatives and then in situ into the iodides by a Finkelstein reaction. The speed of ring formation yielding the tertiary amine *112* therefore is increased because of the more pronounced nucleofugicity of the iodides. As alkylations of this type depend on the concentration of the iodide anions and as the potassium iodide, formed in this reaction, has a low solubility in iso-pentanol, a double molar amount of NaI is necessary in relation to the amount of K_2CO_3. Under these conditions, the di-ansa compound *112* was formed in 60% yield [151b]. *112* was isolated in only 44% [151c, d], when a molar ratio K_2CO_3/NaI of 1:0.8 was employed.

In some cases, Na_2CO_3 has turned out to be more successful than K_2CO_3, especially with ansa compounds of type *112* with certain substituents [148d].

The second intermolecular C—N-ring closure leads to the prae-[3]catenanes *115a,b* [151b]. For this purpose, the OH functions in compound *112* were transformed to yield the dibromo compound *113* and the bis-sulfone amide *114*. These were cyclized as educts in equimolar amounts of boiling DMF using K_2CO_3 as a base as described above. (It may be pointed to the fact here that educt *114* is not an aromatic amine as above.) In this 1C-DP-reaction [6a] two isomeric prae-[3]catenanes *115a* and *115b* are isolated in a total yield of 38% [151b].

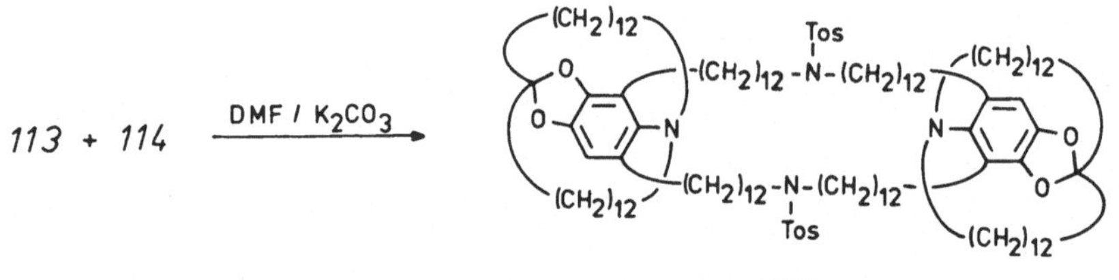

$$113 + 114 \xrightarrow{\text{DMF / } K_2CO_3}$$

115a

+

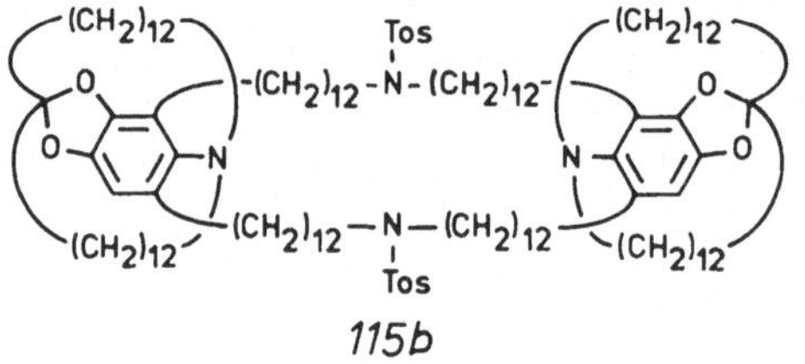

115b

Splitting the bonds in *115a,b* between the aromatic rings and the bridgehead atoms yields the [3]catenanes [151b].

The cyclizations of bis(bromomethyl)-m-xylene (*39*) and N,N'-ditosyl-m-phenylene diamine (*116*) show surprising results [152]. Different from the product formation of the corresponding sulfur compounds (Sects. II.1.3.2, II.1.4.2) [64, 153], which can be completely shifted to the side of the monomeric dithiaphanes through high dilution, in the case of the aza compounds discussed here, only the dimeric tetraazaphane *117* is formed even under very high dilution [152, 154]:

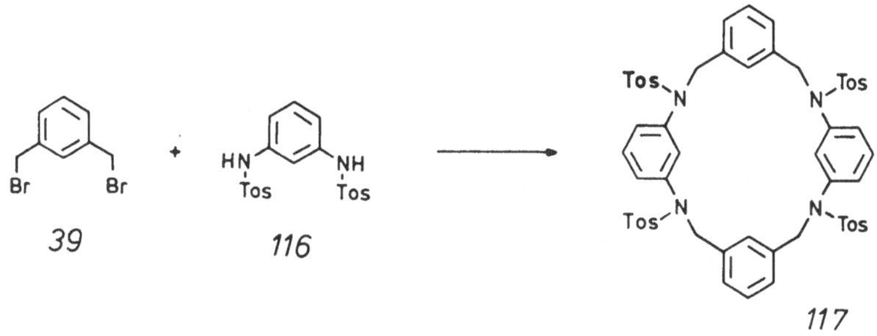

II.4 Synthesis of Medio- and Macrocyclic Compounds by C—C-Bond Formation

For the synthesis of medio- and macrocycles there seem to remain only few well examined possibilities to form C—C-bonds by nucleophilic substitution at the saturated C-atoms with respect to the numerous known general C—C-coupling methods: a) by alkylation of malonic esters and b) by Wurtz synthesis and its variants. Usually halogeno compounds represent the leaving group bearing components.

II.4.1 C—C-Bond Formation by Alkylation of Malonic Esters

II.4.1.1 Cycloalkanes

Whereas 3- to 7-membered rings are easily accessible by alkylation of malonic esters with 1,ω-dibromo alkanes [155], little is known of the syntheses of macrocycles according to this method [156].

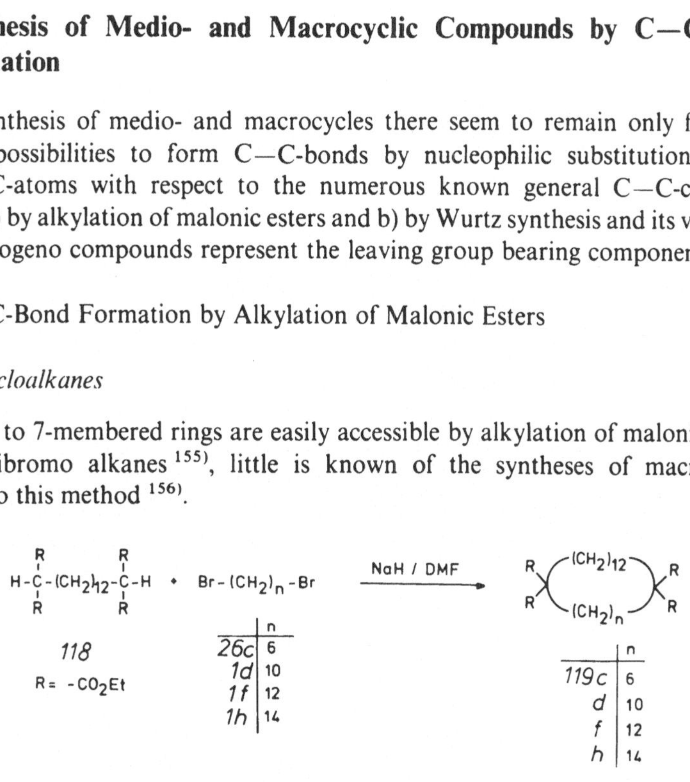

Experimental procedure for the preparation of the 20-, 24-, 26- and 28-membered tetraesters 119c, d, f, h [156]:

starting components: tetraester *118* (0.01 mole) and 1,ω-dibromo alkanes *1d, 1f, 1h, 26c* (0.01 mole) in 250 ml of DMF

reaction type: 1C-DP [6a]

reaction medium: DMF (750 ml) containing NaH (0.15 mole)

reaction temperature: 42–44 °C

time of addition: 24 hrs

additional reaction time: 12 hrs

yields: *119c*: 28%, *119d*: 45%, *119f*: 39%, *119h*: 27%

Although sodium hydride decomposes DMF slowly at higher temperature [157a−c], and also reacts with halogeno alkanes in this solvent, cyclizations are, nevertheless, possible under these conditions. This may be due to the high reaction speeds of alkylations of dialkyl malonates in DMF [157d)], which are quicker than undesired side reactions.

Compared to the above described 1C-DP reaction [6a)], the cyclization starting with diethyl malonate (*120*) itself shall be sketched out:

	n			n			n
1d	10		*121d*	10		*122f*	12
1f	12		*119f*	12		*122h*	14
1h	14		*121h*	14			

$R = -\overset{O}{\underset{O-Et}{C}}$

Under experimental conditions analogical to those described above, two ring closure products *119f*, *121d,h* and *122f,h* are obtained here [156)]. The yields seem to depend on the chain lengths of the dibromides due to transanular effects. Only the 1,12-dibromododecane (*1f*) and the 1,14-dibromotetradecane (*1h*) are in a position to form the diester rings *122f* and *122h*. Whereas *122f*, containing 13 ring members, is obtained only in 1 % yield, the two ring members more containing ring compound *122h* is isolated in 12 % yield. The corresponding diester ring with the decamethylene chain (n = 10) was not indicated [156)].

The ring closure reaction of *120* with 1,12-dibromododecane (*1f*) delivered *119f* in 18 % yield, whereas the same macrocycle *119f* — the cyclization product from *118* and *1f* — occured in 39 % yield. This yield decrease seems to be due to the higher numbers of reaction partners, which take part in the ring closure reaction: two starting with *118* and *1f*, but four in the ring closure of *120* and *1f* [156)].

By variation of several dilution principle conditions, the result regarding the yield of *119f* starting with *120* could not be improved. Using another solvent, e.g. a mixture of THF/DMF (9:1), the yield decreased even more from 18 % to 9 %. Also, the use of LiH as a base in DMF decreases the amount of *119f* to 8 %.

Compared to the yields of *119d, h*, the amounts of *121d* (17 %) and *121h* (11 %) also are quite low [156)].

Macrocycle *125* with a ring width of 27 C-atoms, containing three diethyl malonate units, was recently prepared [158)]. No details of the cyclization are mentioned by the authors, apart from the note that the cyclization was carried out under dilution principle conditions. One can just result from the reaction scheme noted by the authors that *123* and *124* in THF with NaH as the base yield the hexaester *125* in a remarkable yield of 46 % [158)].

Following the same procedure the polyoxa cycle *126*, likewise containing 27 ring atoms, is isolated in only 17.5% yield [158].

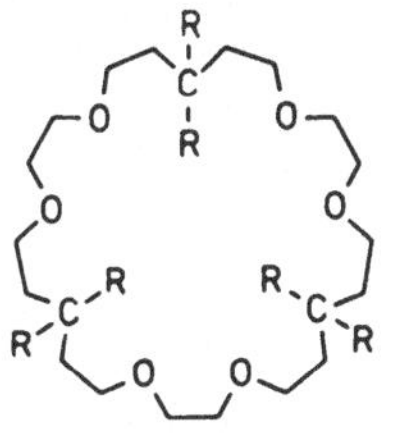

126

R = -CO$_2$Et

Starting with only one cyclization component, an ω-bromoalkyl malonate, under high dilution conditions intra- and intermolecular products are formed depending on the chain length of the educts [159]. DMSO has shown to be especially suited for this type of cyclizing alkylation. [18]crown-6 has been used successfully to increase the nucleophilic power of the anion of the base [159].

II.4.1.2 Phane Esters

Up to some years ago the alkylation of malonic esters was one of the most important methods for the synthesis of [3.3]phanes. Today, such phane hydrocarbons can be prepared in a shorter way by desulfurisation of [4.4]thiaphanes (cf. Sects. II.1.1.3, II.1.3.2, II.1.4.2) by pyrolysis [37] or photolysis [36].

Also, the necessity to employ strong bases in excess, limits this classical direct C—C-coupling method to educts and products that are insensitive to bases. As a rule therefore the synthesis of [3.3]carbaphanes is favourably achieved by the thiaphane method.

Apart from one exception, the syntheses of the [3.3]carbaphane esters succeeded only in using monoalkylated dialkyl malonates as the starting material. The not alkylated dialkyl malonate could only be used successfully for the preparation of the 2,2-dicarbethoxy[3.2]metacyclophane (*128*) and its dimer *129*. Diethyl malonate (*120*) was cyclized with 3,3′-bis(bromomethyl)bibenzyl (*127*) in boiling xylene with excess NaH [160].

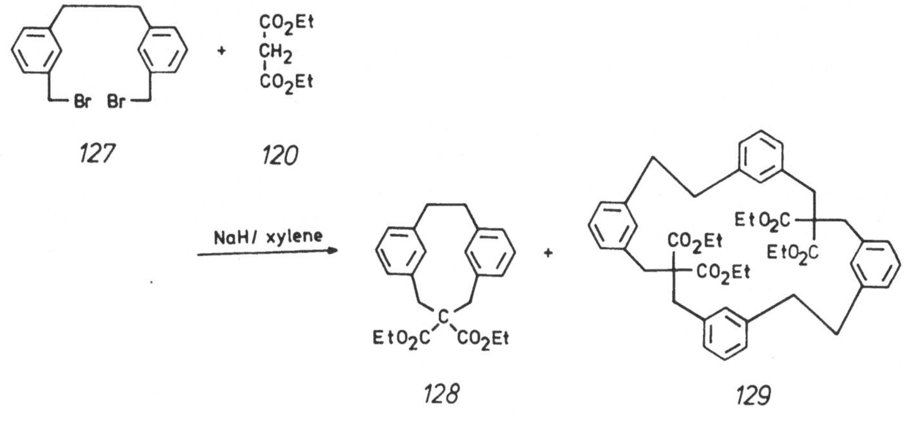

127   *120*

128   *129*

In this 1C-DP-reaction [6a)] the 11-membered diester *128* was obtained in 11 % yield, the dimer *129* was formed in higher yield of 16 % [160)]. The relatively good result of 11 % for the mediocyclic ester *128* with a ring width of 11 atoms, lying in the so-called yield minimum region [4)], may be referred to the rigid group principle [3, 11a, 12)].

Experiments to synthesize a 2,2,11,11-tetracarbethoxy[3.3]metacyclophane (*130*) [161,] by alkylation of diethyl malonate (*120*) with 1,3-bis(bromomethyl)benzene (*39*) in a 2C-DP-reaction [1a)] in dioxane with NaH, were not successful.

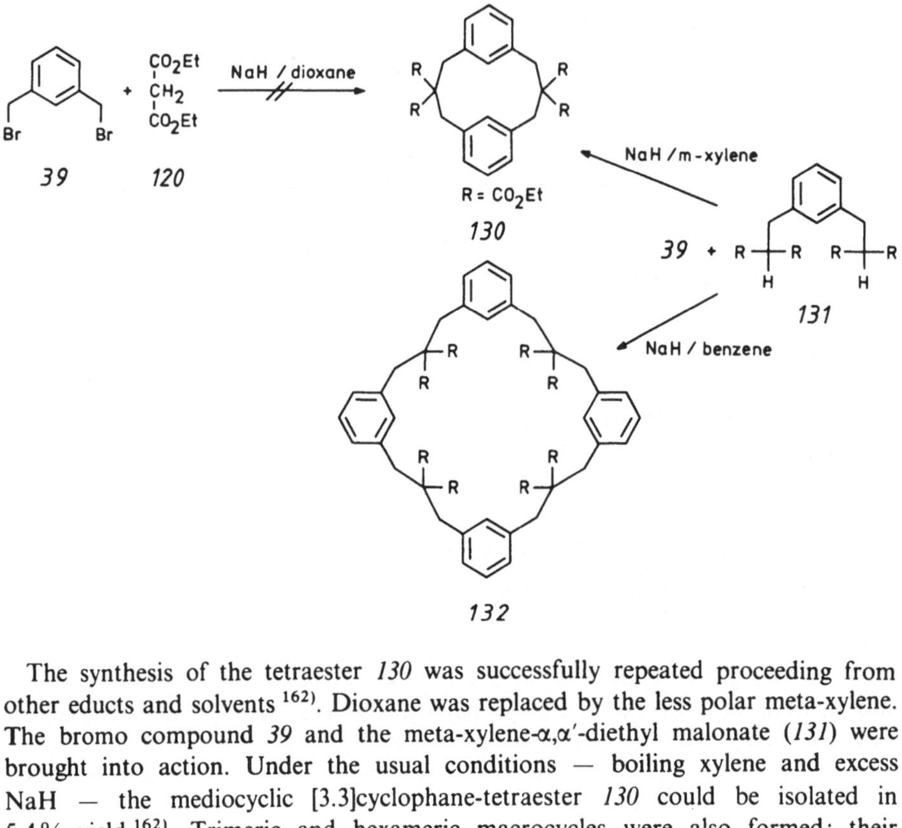

The synthesis of the tetraester *130* was successfully repeated proceeding from other educts and solvents [162)]. Dioxane was replaced by the less polar meta-xylene. The bromo compound *39* and the meta-xylene-α,α'-diethyl malonate (*131*) were brought into action. Under the usual conditions — boiling xylene and excess NaH — the mediocyclic [3.3]cyclophane-tetraester *130* could be isolated in 5.4 % yield [162)]. Trimeric and hexameric macrocycles were also formed; their percent amount was not reported.

In cyclization reactions of this type, the solvent plays an important role: If both components *39* and *131* (20.0 mmole) are dissolved in 250 ml of benzene each [161)], and dropped into a suspension of 0.5 mole of NaH in 1.5 l of benzene, using two precision dropping funnels [6b, 23)], only the tetrameric octaester *132* is obtained in 7 % yield [161)]. No other cyclic product could be indicated under these conditions.

The formation of the cyclic trimer of *130*, the 2,2,11,11,20,20-hexacarbethoxy-[3.3.3]metacyclophane, which occurs as a byproduct in 2 % yield in the synthesis of *131* starting with the acyclic compounds *120* and *39* is remarkable [161, 162)]. The formation of the hexaester is even more remarkable, as it was obtained without the help of the dilution principle in strong polar solvents (ether, ethanol; cf. lit. [1a)]). In general less polar or unpolar solvents are qualified for C—C-coupling reactions.

The advantages of the synthesis of carbaphanes passing the thiaphane route mentioned in the introduction shall be made plain here only using one example: the cyclizations, yielding the tetraester *130* [162] and 2,13-dithia[4.4]metacyclophane (*136*) (Sect. II.1.4.2) [98] and their transformation to [3.3]metacyclophane (*135*) [98, 162], shall be compared:

C—C-coupling:

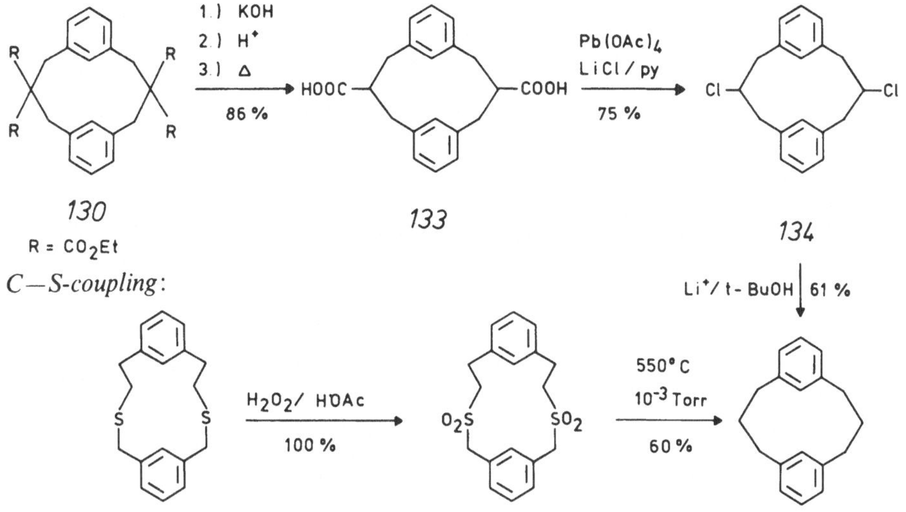

$$R = CO_2Et$$

130   133   134

C—S-coupling:

136   137   135

The key steps for the synthesis of *135* are the cyclization reactions yielding *130* and *136*. Herein lies the essential difference between C—C- and C—S-bond formations: Whereas *130*, as mentioned above, is formed in only 5.4% yield [162], the thiaphane *136* according to the thiol procedure (Sect. II.1.4.2) is isolated in 31% yield [98]. Furthermore the formation of the carbaphane *135* beginning with *130* proceeds via two steps [162], whereas beginning with the thiaphane *136*, the metacyclophane *135* is obtained in only one single step, passing only one single intermediate, the sulfone *137* [98]. The oxidation yielding the sulfone *137* proceeds quantitatively as usual. The removal of the four ester functions in *130* can be achieved with relatively good yields (cf. reaction equation), but they do not reach the yields of the sulfide oxidation.

A row of other [3.3]carbaphanes was synthesized by the alkylation of malonic esters [164a–c], but their preparation also could be improved using the [4.4]thiaphane method [93f, 96–98]. These [3.3]phanes mainly have been studied under the aspect of intramolecular charge transfer interactions.

II.4.2 C—C-Bond Formation by Wurtz-Reaction and its Modifications

Although intensive studies have been devoted to the mechanisms of the Wurtz reaction and its modifications, firm conclusions are still lacking, because the mechanisms may vary depending on the metal used, the substituents, the catalysts, if any, and other reaction conditions. Two basic pathways can be envisioned: a nucleophilic substitution process or a free radical mechanism. For simplification we treat this reaction type under topic II.

II.4.2.1 Phanes

The Wurtz C—C-coupling and its variations offer another possibility for the formation of bonds between saturated C-atoms. They have been used for the synthesis of numerous cycloalkanes [165] and phanes [33b]. To illustrate the differences between the several Wurtz types, this section restricts to the preparation of phane systems according to the dilution principle. The following three reaction types are compared: a) the original Wurtz synthesis, b) the Müller/Röscheisen modification, and c) cyclizations with organolithium compounds.

a) Wurtz synthesis:

In the historical studies of Wurtz [166] the coupling reactions of acyclic mono-halogeno alkanes with sodium to yield open chained hydrocarbons are described. Some years later this method was transferred to ring closure reactions. One example under numerous other phane system syntheses [167–171] shall be described here to show the characteristics of this method: the intermolecular 1C-DP-reaction [6a] of 1,2-bis(bromomethyl)benzene (*82*) to yield [2.2]orthocyclophane *138* [171]:

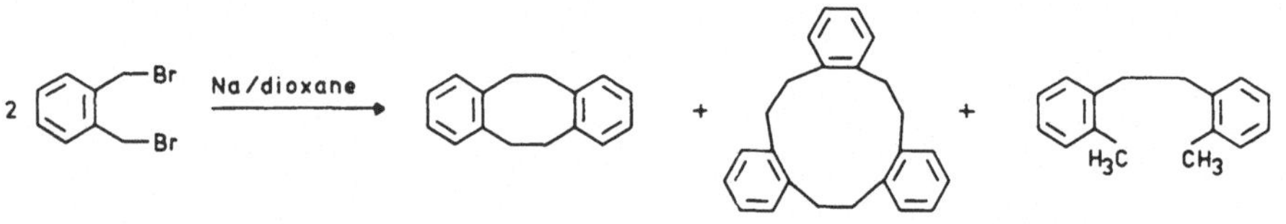

Experimental procedure for 138 [171]: starting component: 1,2-bis(bromomethyl)benzene (*82*) (0.25 mole) in 250 ml of dioxane reaction type: IC-DP [6a]; reaction medium: dioxane (250 ml) containing sodium (0.61 gat); reaction temperature: boiling dioxane; time of addition: 24 hrs; additional reaction time: 4 hrs; yield: 30% of *138*; byproducts: minor amounts of trimer *139* and 1,2-bis(2-tolyl)ethane (*140*).

In the same study it was inquired the yield of *138* depending on the concentration of *82* [171] in dioxane: the highest yield of *138* (46%, 0.12 mole of *82*) was obtained without the application of the dilution principle (cf. ref. [1a]) in extremely diluted dioxane solution (1.32 l) at a reaction time of 48 hrs. Nevertheless, from an economical point of view, the dilution principle cyclization with the lower yield of 30% of *138* was preferred, because the total reaction time could be decreased to 28 hrs and the dioxane volume to only 500 ml.

At a higher concentration of *82* in dioxane the yields strongly decrease as far as 6% [167].

Not only the above mentioned bimolecular [171], but also the monomolecular cyclizations under C—C-coupling [168, 170] were performed according to the 1C-DP-procedure [6a]. Furthermore dibromo compounds as starting components are characteristic. Most often benzylic dibromo compounds [167–171] were employed owing to their reactivity. Boiling solvents of low polarity like dioxane and xylene were applied most often [167–171]. Bromobenzene or para-bromoanisol are frequently added as catalysts to accelerate the C—C-ring coupling [169].

Rarely dilution principle cyclization according to the Wurtz method proceed with such high yields as in the preparation of *138* [171]. The rigid group principle [3, 11a, 12] is made responsible for this good result. In general, Wurtz syntheses are characterized by a high amount of unwanted polycondensed products.

b) The Müller/Röscheisen variant:

To circumvent the heterogeneous Wurtz synthesis [172)] with its polycondensed by-products. Müller and Röscheisen tried to dissolve the metal [172)] by forming addition compounds of alkali metal with aromatic substituted ethenes and dienes. The most accepted addition compound, also applied in the Müller/Röscheisen C—C-coupling reaction, is tetraphenyl ethene (TPE).

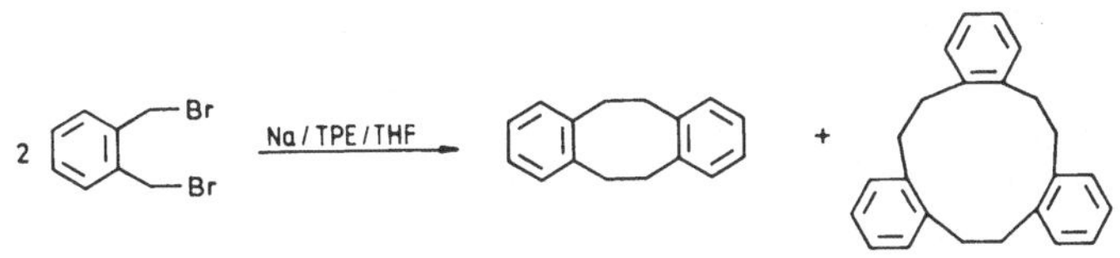

Experimental procedure for 138 and 139 starting with 82 according to the Müller/Röscheisen method [172)]:

To a solution of 150 ml THF, 4.0 mmole TPE and 0.40 gat sodium in the reaction flask a solution of 80.0 mmole of 1,2-bis(bromomethyl)benzene (82) in 100 ml of THF is slowly added dropwise under stirring at –80 °C. After a reaction time of 10 hrs the mixture is worked up: the total yield is 75.5%, composed of 138 (40%) and 139 (35.5%) [172)].

The amount of 138 is as high as in the original Wurtz-reaction, but the total amount of cyclic products is remarkably enhanced and the amount of undesired byproducts is significantly repressed.

An essential distinction between the Müller/Röscheisen variant and the original Wurtz-reactions is the low cyclization temperature of -50 to -80 °C. All ring closure reactions with metal addition compounds are carried out in THF [172–179)].

An additional comparison of the Wurtz- and Müller/Röscheisen cyclizations is offered by the following reactions [170)]:

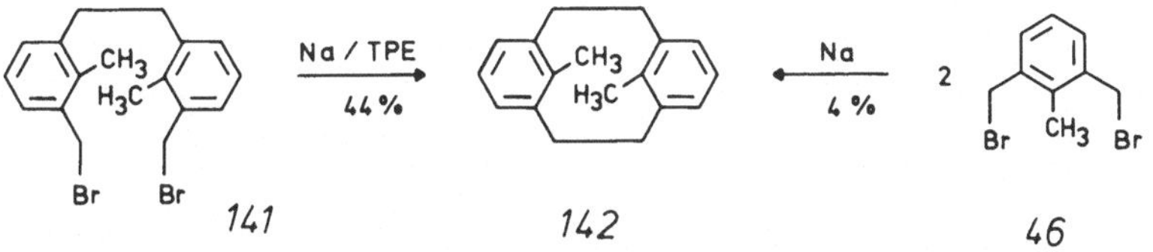

The monomolecular dilution principle cyclization beginning with 141 was achieved according to the Müller/Röscheisen modification at -50 °C in THF. The 8,16-dimethyl[2.2]metacyclophane (142) was obtained in 44% yield, whereas in the intermolecular reaction with sodium in dioxane proceeding from 46, 142 could be isolated in only 4% yield [170)]. The yield difference of 40% is due not only to the homogeneous phase in the Müller/Röscheisen variant, but also to the fact that in the cyclization 46 → 142 two C—C-bridges must be closed, whereas in the reaction 141 → 142 only one C—C-bond is formed.

In the context of the syntheses of substituted [2.2]carbacyclophanes such cyclizations are of interest which begin with different educts and yield the same product [173)].

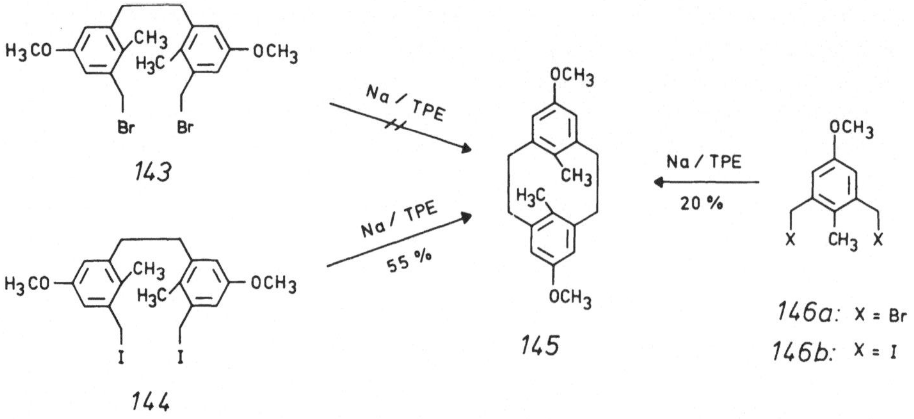

Experiments to synthesize the 8,16-dimethyl-5,13-dimethoxy[2.2]metacyclophane (*145*) by intramolecular cyclization of the bromo derivative *143* with sodium and TPE in THF at −80 °C, were unsuccessful. Only the corresponding iodide *144*, prepared from the dibromide *143* by Finkelstein reaction, formed the cyclophane *145* under otherwise analogous conditions in 55% yield. In contrast to that the bimolecular ring closure reaction of *146* under the same conditions lead to 20% of the 10-membered ring compound *145* independent of the type of the halogeno substituent in the starting component [173].

c) Cyclizations using organo-lithium compounds:

A further alternative to form C—C-bonds is the use of phenyl-lithium or n-butyllithium as coupling reagents. The advantage of lithium organic compounds on the one hand is that cyclizations can proceed in homogenous solution as in the Müller/Röscheisen variant, and on the other hand that they exhibit a higher selectivity than sodium in nucleophilic substitution reactions (cf. Sect. II.4.2). The latter attribute is decisive for the decrease of side product formation, especially of polycondensed material.

The solutions of the lithium organic compounds are usually prepared directly in advance of the cyclizations by the reaction of elementary lithium with bromobenzene or n-bromobutane resp. in ether, THF or a mixture of ether and benzene. The concentration of the lithium organic compounds in solution is determined titrimetrically [180].

As the syntheses of phane systems with lithium organic compounds proceed according to the same principle, we only give one experimental procedure, and only pointing to the literature dealing with dilution principle syntheses of further phanes [181-186].

Experimental procedure for the preparation of [2.2]metacyclophane (147) [181]:

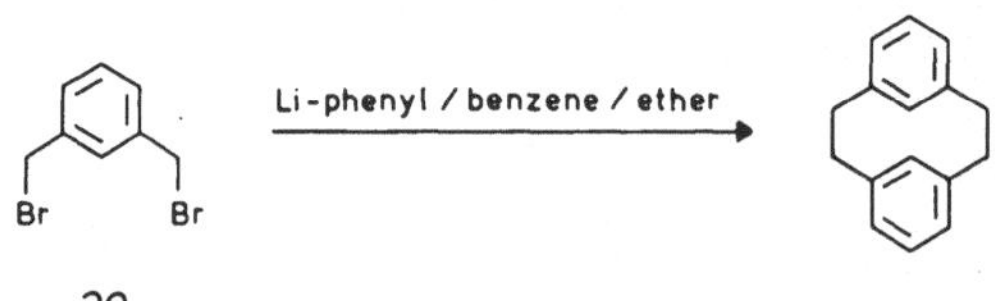

starting component: 1,3-bis(bromomethyl)benzene (*39*) (39.0 mmole) in 600 ml of benzene
reaction type: 1C-DP [6a)]
reaction medium: benzene/ether (1:1.5) (250 ml) containing phenyllithium prepared from bromobenzene (43.0 mmole) in 100 ml of benzene and lithium (0.1 gat) in 150 ml of ether
reaction temperature: boiling solvent
time of addition: 6 hrs
additional reaction time: 2 hrs
yield: 39% of *147*

In this method of C—C-coupling only the temperature and the type of addition are varied. The cyclizations are performed either at room temperature or in slightly warmed or in boiling solvents. As all ring closure reactions are accomplished according to the 1C-DP-method [6a)], there are only two possibilities of addition: either the solution of the lithium organic compound is given into the solution of the educts [182, 183, 186)], or the starting compounds are added dropwise to the solution containing the organo-lithium compound [181, 184, 185)].

For reasons mentioned in Section II.4.1.2, today there are more productive procedures for the preparation of carbaphanes, e.g. the thiolate and the thiol method (cf. Sects. II.1.3.2, II.1.4.2) with subsequent desulfurisation. In this connection the intermolecular cyclization of the tetrahydrodibenzoanthracen derivative *148* to yield the macrocycle *149* shall be mentioned. Up to some years ago, this ring closure yielding *149* was carried out by direct C—C-coupling with phenyllithium and the phane *149* was obtained in only 1.5% yield [185)]. Later, the macrocycle *149* was prepared more effectively in 60% yield [96b)] passing the cyclic thiaphane *151* and successive photochemical extrusion of the sulfur atoms. The thiaphane *151* thereby was obtained from *148* and *150* under consideration of the dilution principle according to the thiol method (Sect. II.1.4.2) in 55% yield [96b)].

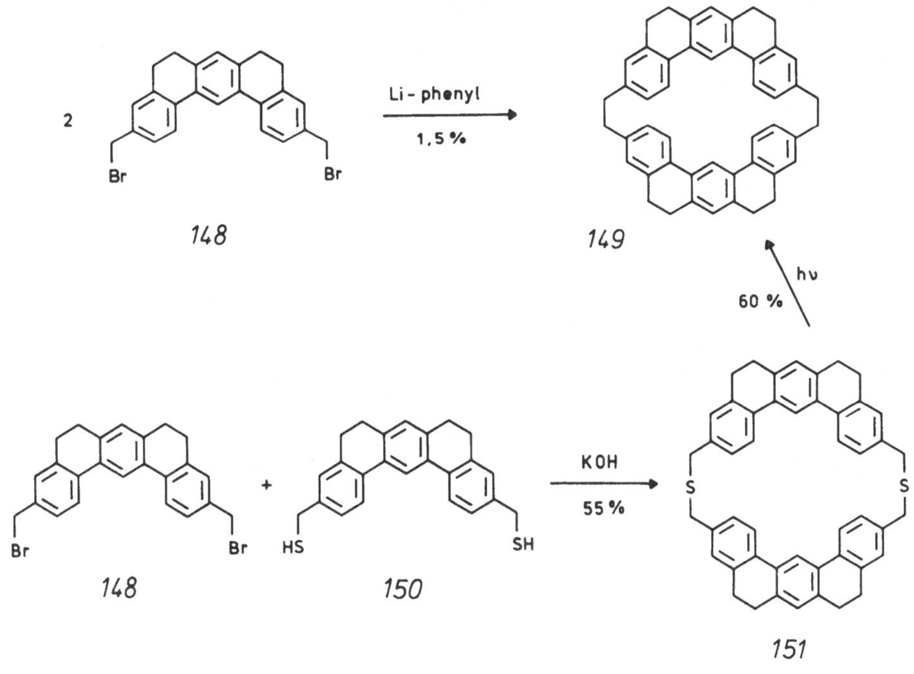

148   *149*

148   *150*   *151*

III Nucleophilic Substitutions at Unsaturated C-Atoms

Carbonyl and carboxyl derivatives as well as nitriles and others are used as electrophilic C-components in dilution principle reactions leading to medio- and macrocyclic systems by nucleophilic substitution at unsaturated C-atoms.

In contrast to the numerous methods of C—S-bond formation by nucleophilic substitution at saturated C-atoms (Sects. II.1.1–II.1.4) the C—S-coupling at unsaturated C-atoms could rarely be utilized for the synthesis of medium- and many-membered rings up to this day [187].

III.1 Synthesis of Medio- and Macrocyclic Compounds by Formation of Ester Bonds

The formation of medio- and macrocyclic systems with one or more ester functions is important in many ways: a) for the synthesis of macrolides [188, 189], b) for the preparation of model substances [29a, 119] with complexation properties similar to those of the macrolides, and c) from a theoretical point of view, e.g. for the study of influences of the ester functions on the ring closure reaction [1a, d, 12c–e, 119, 190] or the steric interactions in the interior of oligoester ring systems [34a, 72a, 152, 191].

The formation of an ester as a rule proceeds with comparatively low speed [4, 192]. It is therefore understandable that as well the intermolecular ring closure reaction as the formation of acyclic side products prevail the intramolecular ester formation. The preparation of medio- and macrocyclic lactones, the intramolecular esters of ω-hydroxycarbonic acids therefore demands the activation of the acid and/or hydroxylic function [189, 192, 193]. As there were several reports on the numerous methods for the activation of ω-hydroxycarbonic acids for the synthesis of lactones [119, 189, 193, 194], this review section shall deal with lactones and oligolactones together, under retention of the above used classification into the aliphatic ring systems, crown ether- and phane systems under the aspect of dilution reactions.

III.1.1 Ester Bond Formation by Reaction of Carboxylic Acid Derivatives with Hydroxy Compounds

III.1.1.1 Cyclic Aliphatic Esters

The main interest in the preparation of cyclic aliphatic esters focusses on the synthesis of naturally occuring aliphatic macrolides [195]. The final reaction step consists in a cyclizing formation of ester bonds regardless of the number of ester functions already present in the open chained starting component. As a common aim of these reactions it is aspired to favour intramolecular ring closures and at the same time to depress intermolecular cyclization reactions. To reach this aim, several methods are applied which are suited for dilution principle reactions:
a) the acid catalysed direct esterification [196], b) the formation of esters by activation of the carboxyl function [197], e.g. through formation of acid chlorides or of anhydrides [198], through formation of N-acylimidazoles [199, 200], c) by activation of the carboxyl- and hydroxyl function with 2-pyridine thiol esters [192].

One of the first and most simple macrolides, its isolation and structure elucidation was accomplished by Kershbaum [201] in 1927, is the 16-membered lactone exaltolide *153*. By the aid of this simple lactone, the first two of the above mentioned methods shall be described in some detail:

HO—(CH$_2$)$_{14}$—C$\lessgtr^O_{OH}$ ⟶   (CH$_2$)$_{14}$ C=O

152   *153*

Stoll and Rouvé [196a] carried out two experiments for the synthesis of *153* beginning with 14-hydroxytetradecane-1-carboxylic acid (*152*).

Experimental procedure for exaltolide (153) [196a]:
starting component: 14-hydroxytetradecane-1-carboxylic acid (*152*) (38.7 mmole)
reaction type: 1C-DP [6a]
reaction medium: dry benzene (10 l) containing dry benzene-sulfonic acid (18.3 mmole)
reaction temperature: boiling solvent
time of addition: 6 days
additional reaction time: none
yield: 87% of *153* besides 4% of polycondensed products

In a second dilution principle experiment the following parameters were varied: a) the time of addition was shortened from 6 days to 3 hrs, b) more than the threefold amount of benzene sulfonic acid was used. The calculated amount of water could be collected already after 33 hrs instead of the 6 days. Surprisingly, the shortening of time didn't result in a significant decrease in the yield of the lactone *153*: instead of 87% of *153* besides 10% of polycondensed byproduct 76.5% were obtained in the second experiment. The excess of benzene sulfonic acid is made responsable for the good yield in despite of the shorter reaction time.

The high dilution reaction *152* → *153* proceeds with astonishing high yields. This might be explained apart from the effectiveness of the dilution principle by the number of 16 ring members, which lie above the yield minimum [4] observed in the medium ring region.

A mild and effective method for esterification consists in the simultaneous activation of the carboxyl and hydroxyl function [192]. This double activation is made possible by the reagent 2,2'-dipyridyl disulfide which forms the thioester. The reaction steps are depicted below, including the intermediate ω-hydroxy alkane thiol ester *155* [192, 193].

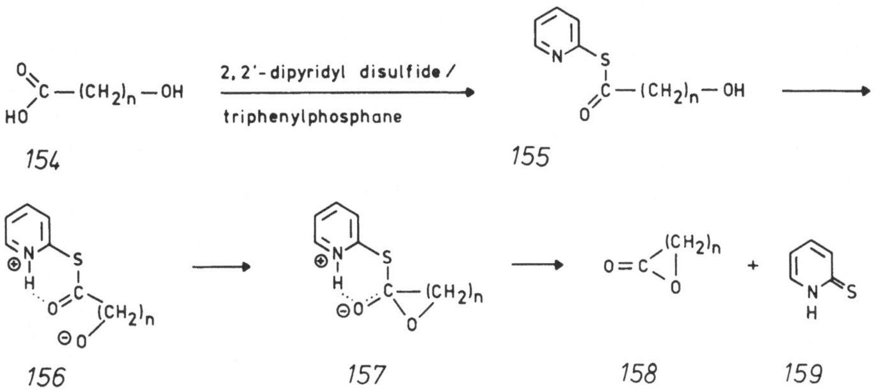

2,2'-dipyridyl disulfide /
triphenylphosphane

154   *155*

156   *157*   *158*   *159*

47

Before this method could be applied for the synthesis of naturally occuring macrolides, its effectiveness was tested for such ω-hydroxy carboxylic acids, the ring width of which after cyclization lies in the critical range of 9–12 atoms. It was especially tried thereby to suppress the formation of dimeric compounds:

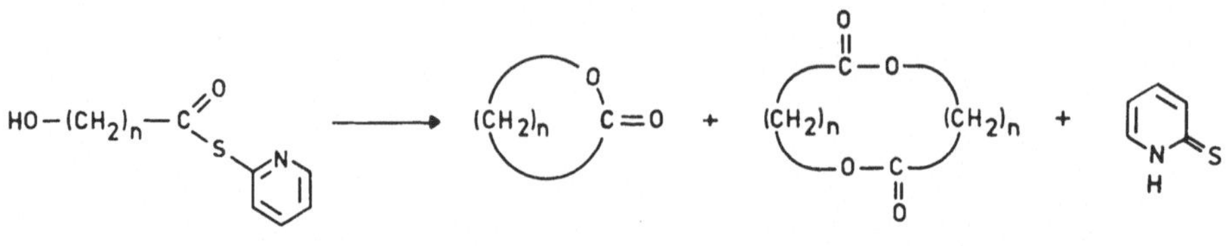

155a–f

n = 5, 7, 10, 11, 12, 14

158a–e

153

160a–f

159

Experimental procedure for the lactones 158a–e, 153 and the cyclic diesters 160a–f [192, 193]:
starting component: 2-pyridine thiol esters *155a–f* prepared from the corresponding hydroxy acids
        154a–e, 152 (0.5 mmole), 2,2'-dipyridyl disulfide (0.75 mmole) and triphenyl-
        phosphane (0.75 mmole) in xylene
reaction type: 1C-DP [6a]
reaction medium: xylene (100 ml)
reaction temperature: boiling xylene
time of addition: 15 hrs
additional reaction time: 10 hrs (20 hrs for *155c*, 30 hrs for *155b*)
yield: see Table 1

Table 1. Yields of mono- and dimeric lactones according to the dipyridyl disulfide method [192]

Cpd. nr.	n	ring width	yield (%)	Cpd. nr.	n	ring width	yield (%)
158a	5	7	71	*160a*	5	14	7
158b	7	9	8	*160b*	7	18	41
158c	10	12	47	*160c*	10	24	30
158d	11	13	66	*160d*	11	26	7
158e	12	14	68	*160e*	12	28	6
153	14	16	80	*160f*	14	32	5

A comparison of the yields shows that method c) [192] (cf. p. 46) well suits the preparation of lactones. With the exception of the strained 9-membered ring system *158b* the intramolecular cyclizations yielding *158a, 158c–e, 153*, are preferred compared with the intermolecular ring closure reaction yielding the diesters *160a, c–f*. Only the percentual amount of triesters of the acid *152* was published, although the trimers of all 6ω-hydroxy carboxylic acid thioesters *155a–e* were detected. The 48-membered, aliphatic macrocycle containing three ester functions was isolated in the low yield of 1%.

This mild procedure for esterification was transferred successfully to the synthesis of macrolides: (±) zearalenone was synthesized for example according to this method which is discussed in Section III.1.1.3 [193].

To sum up: independent of the method used, all aliphatic lactones mentioned above were synthesized applying a IC-DP-reaction [6a] in unpolar, lipophilic solvents.

The reaction temperature in general is the boiling temperature of the corresponding solvent. The addition times and reaction times vary according to the esterification method and may range from 6 days to 3 hrs. The yield minimum [4] is found in the region of the 9–12-membered aliphatic ring systems as expected.

III.1.1.2 Crown Ether Esters

In contrast to the syntheses of aliphatic esters (cf. III.1.1.1), crown ether ester syntheses usually don't aim at the syntheses of naturally occurring macrolides in the first place [29a, 119]. Crown ether esters were designed more as model substances to study properties such as complex stability, cation selectivity and permeability through membranes [29a, 119, 202, 203]. There were mainly prepared cyclic crown ether lactones possessing more than one ester function [119], the cyclization of which does not demand a special activation, which was generally used for the preparation of mono lactones [192, 193, 196–200] (cf. III.1.1.1). As a rule, carboxylic acid chlorides and in some cases the carboxylic acids themselves were esterified with glycols (cf. III.1.1.1).

Under the large number of known polyoxa esters only those examples are reported here in more detail, which are prepared using characteristic high dilution conditions. The lactonisation of terephthalic acid dichloride (*161*) with tetraethylene glycol (*162*) for example exhibits general valid dilution parameters [204, 205]. Also the product steering in dependence of the dilution ratio can be shown here impressively.

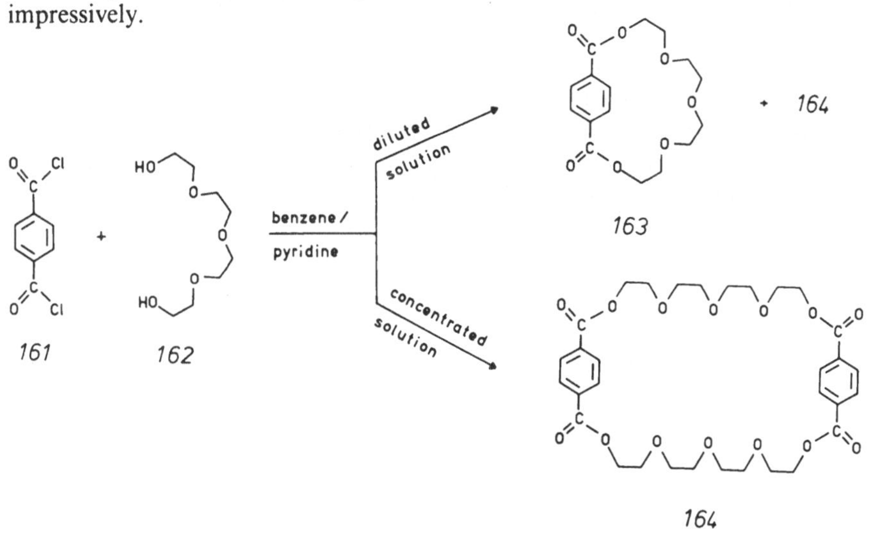

Experimental procedure for 163 and 164 according to the dilution principle [205]:
starting components: a) terephthalic acid dichloride (*161*) (5.00 mmole) in 250 ml of benzene
  b) tetraethylene glycol (*162*) (5.00 mmole) and pyridine (10.0 mmole) in 250 ml of benzene
reaction type: 2C-DP [6a]
reaction medium: benzene (2 l)
reaction temperature: boiling solvent
time of addition: 22 hrs
additional reaction time: none
yield: 13% of diester *163*, 2% of tetraester *164*

This cyclization was also accomplished under the following varied dilution parameters [205]: a) the concentration of the educts *161* and *162* was increased, b) the volume of the solvent, submitted in the reaction flask, was cut from 21 to 750 ml benzene, and c) the addition time was shortened from 22 hrs to 12 hrs. Under these conditions no monomeric dilactone *163* was observed, whereas the dimeric tetralactone *164* was isolated in 20% yield [204, 205].

The above mentioned dilution principle cyclizations show several generally valid characteristic features, which are applied in the syntheses of other polyoxa ether compounds:

a) all cyclic crown ether esters of the above mentioned type were prepared in 2C-DP-reactions [6a], b) benzene is most often employed as a solvent [204–212] and c) the cyclizations are carried out either at 45–50 °C or at the boiling temperature of benzene (81.5 °C). There is a connection between the reaction temperature and the cyclization time: the lower the temperature, the longer — understandably — the reaction time must be chosen. This time varies from 6 days to 6 hrs. d) The addition of pyridine as a HCl-binding agent is not absolutely necessary. The pyridine seems to accelerate the cyclization reaction; otherwise longer reaction times were needed. e) The yields vary strongly and depend on many different factors, discussed in the original literature [202–212].

The synthesis of thiolesters [204, 208, 210, 211] in principle doesn't differ from the type of preparation of the oxaesters. As C—S-bonds in the neighbourhood to a trigonal C-atom are sensible against oxidation, the cyclization reaction must be carried out under protective gas. Also at this way of preparation the few thiolesters known up to now only formed in low yields of less than 5% (exception cf. lit [204, 208]).

Of interest in connection with the dilution principle is a crown ether ester that was prepared using two different ways [211]: The cyclization of the glycol *162* with the free acid *166*, as well as the reaction of the glycol *162* with the corresponding acid chloride *165* has been described in some detail.

The 1,4,7,10,13-pentaoxacyclohexadecane-14,16-dione (*167*) was obtained in a 2C-DP-reaction [6a] beginning with malonic acid dichloride (*165*) and tetraethylene glycol (*162*) according to the above described procedure with a modification in the following parameters: a) the total reaction time was 2 days, b) the temperature of the reaction mixture was 50–60 °C, and c) the HCl-gas wasn't bound. Under these conditions, *167* was isolated in 27% yield [211].

If malonic acid (*166*) is esterified directly with the glycol *162* in boiling benzene without regarding the dilution principle (cf. lit. [1a]), using a trace of para-toluene sulfonic acid, one obtains the macrocyclic diester *167* in 1–2% yield only [211]. The formation of other products wasn't mentioned. The large yield difference of 25–26% may not only be explained by the higher carbonyl activity of the acid chloride compared with the free carboxylic acid, but it has also to be taken into account that in the experiment proceeding from free malonic acid (*166*), it was worked according to the "batch-wise" procedure [1a, 211].

III.1.1.3 Phane Esters

The phanes synthesized hitherto which possess one or more ester functions can be classified into two groups, which differ in their object in view: the former is directed to the synthesis of naturally occurring macrolides [189, 193, 213–217], the second is stimulated by the interest in physical properties as e.g. the spatial requirements of substituents in the interior of the macrocycle [34, 72, 191, 218, 219], the physical properties such as melting points etc. in dependence on the ring width [220] or the influence of ring closure facilitating factors [204, 220].

The macrolide zearalenone (*168*), which shows anabolic effects, may be counted to the [12]orthocyclophane lactones:

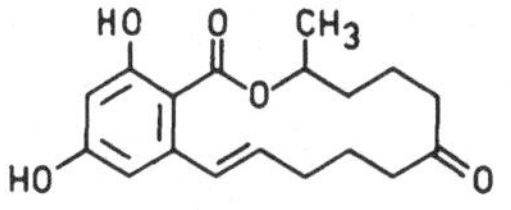

168

Total syntheses for *168* were described by several groups [192, 198, 213, 217]. The starting point for the cyclization yielding *168* is the OH-protected acid *169*, in analogy to the cycloaliphatic lactones. *169* is changed into the corresponding pyridine thiol ester *170* in benzene with 2,2'-dipyridine disulfide and triphenylphosphane at room temperature and then added dropwise to boiling benzene over a period of 10 hrs. This leads to the (±)-zearalenone derivative *171*, which after splitting off the protecting groups yields (±)zearalenone (*168*) in 75% yield [192].

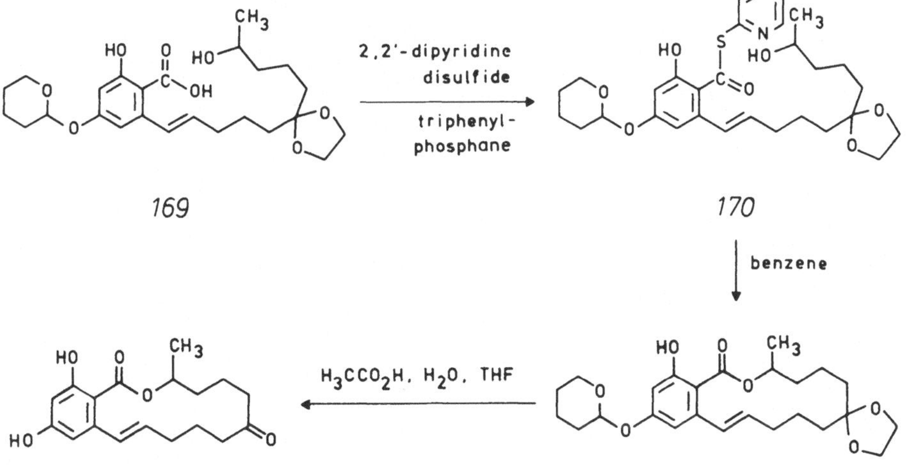

169

2,2'-dipyridine disulfide

triphenyl-phosphane

170

benzene

168

H_3CCO_2H, H_2O, THF

171

51

Another possibility for the activation of carboxylic functions is the formation of an anhydride, most often obtained by the use of trifluoroacetic acid anhydride [198, 213, 217]. According to this method, the zearalenone dimethylether *173* was synthesized [198]. The yield of the lactone *173* depends on the molar ratio of trifluoroacetic acid anhydride and the component *172* [198].

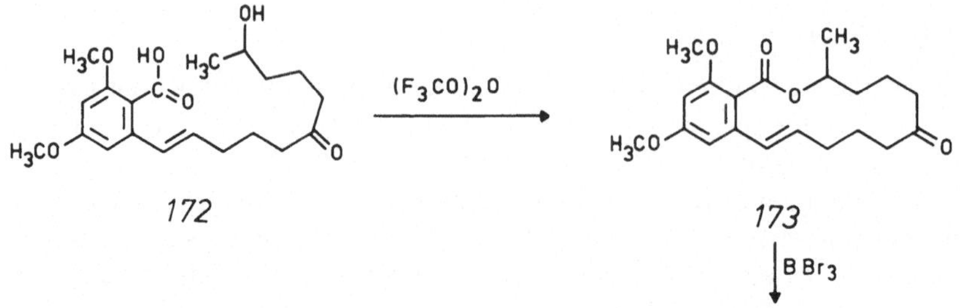

The cyclizations of phane monoesters were all carried out as 1C-DP-reactions [6a] in benzene at a temperature of 6–12 °C.

The acid chloride method is one of the most productive procedures for the preparation of phane esters. All reactions of acid chlorides and alcohols were accomplished using similar dilution principle conditions, for which the following procedure is an example [221]:

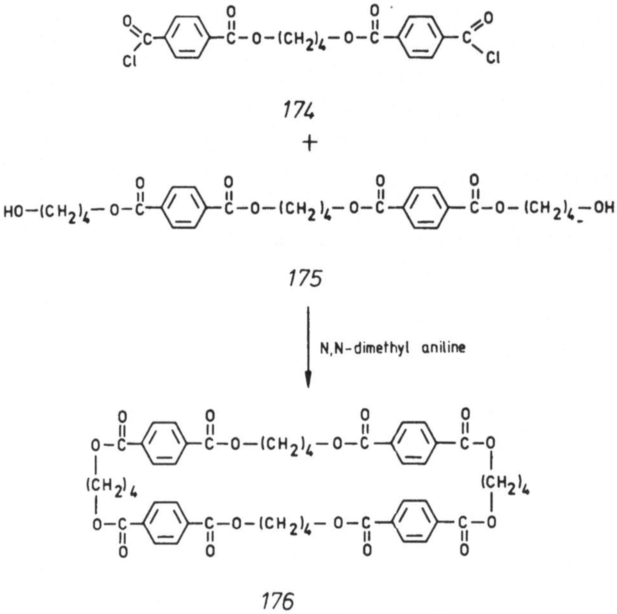

Experimental prodecure for the macrocyclic octaester 176 [221]:
starting components: a) 1,4-bis(4-chloroformylbenzoyloxy)butane (*174*) (4.00 mmole) in 250 ml of o-dichlorobenzene
b) 1,4-bis[4-(4-hydroxybutyloxycarbonyl)benzoyloxy]butane (*175*) (4.00 mmole) and N,N-dimethyl aniline (8.75 mmole) in 10 ml of HMPT and 240 ml of o-dichlorobenzene

reaction type: 2C-DP [6a]
reaction medium: o-dichlorobenzene (3.5 l)
reaction temperature: 150 °C
time of addition: 30 min
additional reaction time: 3.5 hrs
yield: 6.25% of *176*

The following more general valid dilution principle parameters for the synthesis of phane esters bearing more than one ester function, beginning with the acid chloride can be derived: a) all cyclizations are carried out as 2C-DP-reactions [6a], b) lipophilic solvents like benzene or dichlorobenzene are applied to dissolve the educts and as a reaction medium; dioxane and THF were used for the cyclizations, but with remarkable lower yield results [220]; c) it is customary to add proton acceptors like N,N-dimethyl aniline or pyridine, often in little excess.

The rather short addition time of 30 min and also the reaction time of 3.5 hrs only are less characteristic in the above described procedure. These times usually range from several hrs up to some days. Nevertheless, the authors state that longer reaction times ranging from 4.5 hrs to 8 days raise the yield of cyclotetramethylene terephthalate *176* from 7% up to 48% [221].

Magnesium salts added to the cyclization mixture also seem to improve the yields [220]. No explanation is given, but a template effect [11] could be responsible for the favoured ring closure.

Whereas the ratio of monomeric diester and dimeric tetraester in the synthesis of cyclic polyether esters can be steered by the degree of dilution [205] (cf. III.1.1.2), the dilution seems to have little influence on the monomer/dimer product formation in case of the phane esters. This can be illustrated in a comparison of the reactions of pyridine-2,6-dicarboxylic acid dichloride (*177*) with 1,10-decanediol (*178*) and tetraethylene glycol (*162*) resp. to yield the dimeric tetraester phane *179* and the monomeric crown ether ester *180* resp. [204]:

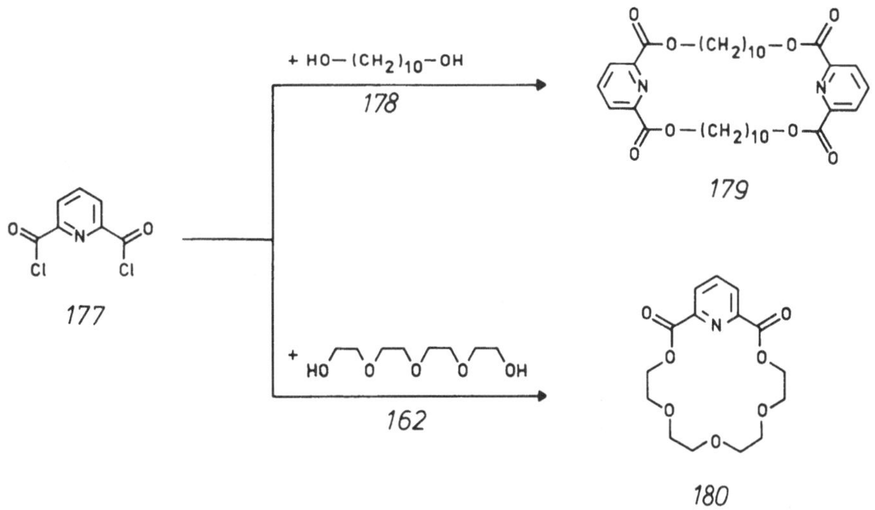

At the same dilution conditions in the phane decamethylene ester synthesis only the dimeric product *179* is isolated in 48% yield, whereas in the polyoxa series only the monomeric cycle *180* is obtained in 70% yield [204]. This different behaviour may indicate a neglected template effect [11] in cyclization reactions including long chained aliphatic diols [204]. For detailed conditions of the synthesis of other phane esters we refer to some surveys [119, 189, 193] and to the original literature [1, 198, 204, 152, 213–221].

III.2 Synthesis of Medio- and Macrocyclic Compounds by Formation of Amide Bonds

Dicarboxylic acid diamides were the first many-membered compounds prepared according to the dilution principle [5, 132]. This ring closure reaction rather ideally meets the conditions required for a cyclization reaction according to the Ruggli/Ziegler dilution principle [4, 5, 132]. These demands are fulfilled with regard to the high speed as well as to the nearly quantitative and distinct course of the reaction (cf. introduction of Sect. II). On ground of these properties the C—N-bond formation at the unsaturated C-atom has won a large scope of application: On the one hand, it was used for the synthesis of naturally occuring lactames, on the other hand, numerous mono- and polycyclic nitrogen containing ligands were developed to study selective cation and anion complexation and to investigate neutral molecule inclusion.

The reduction of the carboxyl function of amides is, as mentioned in Section II.3.1, a productive and universal method for the preparation of medio- and macrocyclic amines.

III.2.1 C—N-Bond Formation by Reaction of Amines with Carboxylic Acid Chlorides and their Derivatives

As in the lactone formation (cf. III.1) in nucleophilic substitutions using amines, carboxylic acids or their activated derivatives are used as electrophilic C-components. Acid chlorides [132, 222], esters [223], as e.g. thiophenyl-, para- and ortho-nitrophenyl- [223] and phosphoric acid esters [224] or the tertiary amide of the thiazolidine-2-thione [225] serve as activated acyl compounds. The most often employed procedure is the acid chloride method, which has become known to be the "Stetter cyclization" [222].

III.2.1.1 Aliphatic Ring Systems

As an example for the broad spectrum of application of the acid chloride method, a reaction shall be chosen which was carried out by Stetter's group [222] and which is characteristic for the method of ring closure reactions. In this amide cyclization reaction 17 macrocyclic diamides ranging from 10- to 21-ring members are synthesized under comparable conditions according to the same method. Therefore, the cyclization of sebacic acid dichloride (*181*) and tetramethylene diamine (*182*) to yield 1,12-diaza-2,11-dioxocyclohexadecane (*183*) shall be quoted [222]:

$$181 \qquad\qquad 182 \qquad\qquad\qquad\qquad\qquad 183$$

Experimental procedure for 183 [222]:

starting components: a) sebacic acid dichloride (*181*) (27.5 mmole) in 500 ml of benzene

b) 1,4-butanediamine (tetramethylene diamine) (*182*) (55.0 mmole) in 500 ml of benzene

reaction type: 2C-DP [6a]
reaction medium: 750 ml of benzene
reaction temperature: 19–23 °C
time of addition: 9 hrs
additional reaction time: none
yield: 74.5% of *183*

Typical marks of an amide ring closure reaction according to the dilution principle are the following: a) all C—N-bonds formed by the acid chloride method are put through as 2C-DP-reactions [6a]; b) unpolar, lipophilic solvents like benzene [222, 226–230], toluene [231] and methanol [232] are used as solvents for the reagents and as the reaction medium; sometimes a mixture of benzene and THF [226] is employed, c) low reaction temperatures seem to be characteristic; they don't exceed 25 °C at any time. Usually the reaction is carried out at 0–4 °C, d) to bind the HCl either the amine functioning as the educt is used in double molar amount or a tertiary amine like triethylamine is chosen [227].

The addition times and reaction times between 2 and 20 hrs depend on the concentration of the educts. The yields turn out generally high, and even in the region of the yield minimum [4] for aliphatic rings they don't range under 20% in any case.

In the same study Stetter and Marx [222] also describe the reduction of numerous cyclic diamides with LiAlH$_4$ yielding the corresponding diamines. These amines are the prerequisite for the successive synthesis of bi- and multicyclic systems [227, 228, 231].

According to this method the tricyclic amine *184* was synthesized by a threefold repetition of the cyclization step yielding the amide and the reduction step yielding the amine [231].

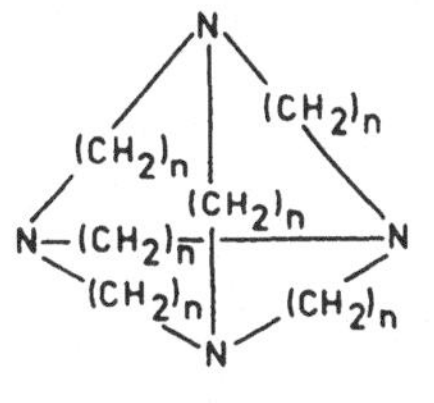

184

With regard to meliorated dilution principle conditions, we can restrict to the monocyclic diamide *187*; in the following cyclization the same parameters were used:

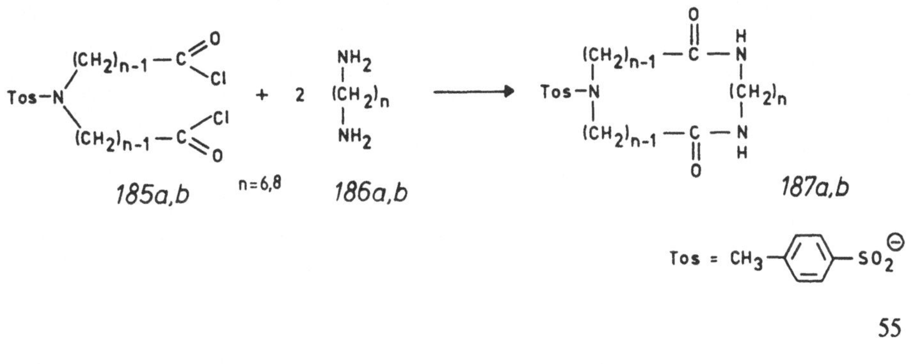

185a,b   n=6,8   *186a,b*   *187a,b*

Tos = CH$_3$—⟨ ⟩—SO$_2^\ominus$

Under strict dilution conditions the reaction of the N-tosyl-protected acid chloride
185a with 1,6-hexanediamine (hexamethylenediamine) (*186a*) — using an apparatus
designed by Stetter [222] — gave the 21-membered diamide *187a* in 30% yield [231].
Benzene hereby was chosen as a solvent at 23 °C.

The yield could even be raised up to 55% [231], when the dilution principle parameters were changed as follows: toluene was chosen as a solvent at a reaction
temperature of 0–2 °C, and the educts were added into the reaction flask by infusion
pumps which can be regulated continuously.

The ring system *187b*, containing 6 ring members more than *187a*, could be obtained
in 40% yield only, even under the optimized conditions. This lower yield reflects the
generally observed decrease of the ring closure yield with increasing ring size [231].

To use the monocyclic diamides *187a, b* as amine components for the succeeding
high dilution ring closure steps, the carbonyl functions must previously be reduced. As
a reagent for the reduction today no longer LiAlH$_4$ [222, 226, 227] is used, but dibo-
rane [228, 231] in THF solution.

The amide cyclization according to Stetter's procedure [222] is also suitable for the
synthesis of natural products. The C—N-ring closure step to yield oncinotine (*188*) [229]
was described, one of three spermidine alkaloids isolated from a Nigerian plant. The
other two alkaloids [230], neooncinotine (*189*) and isooncinotine (*190*) are isomers of
188 and all of them only differ by the connection of spermidine with the C-16 building
block.

188   *189*   *190*

All the three alkaloids are optically active and were isolated in the (*R*)-configuration.
The synthesis of *188* was carried out without consideration of the natural configuration
and yielded the racemic mixture [229].

191   *192*

The open chain educt *191* suitable for the intramolecular ring closure reaction,
containing both the acid chloride and the amine function was obtained in a 14-step
synthesis. Its cyclization to yield the 17-membered lactame *192* was carried out in
benzene with triethylamine as the auxiliary base in 30.9% yield according to the above
described procedure. The lactame *192* contains the skeleton of *188*.

According to the Stetter method [222], the yields in the ring closure step of the intermediates leading to the other two alkaloids *189* and *190* are 75% for the 18-membered N,N'-dibenzyl-neooncinotine and 87% for the N-tosylisooncinotine with a ring member number of 22 atoms [230].

Another activation method useful for a C—N-ring closure is the reaction of 3-acylthiazolidine-2-thiones with amines [225]. The activated educt *196* is prepared by the reaction with thiazolidine-2-thione (*195*) or its thallium salt starting from the carboxylic acid *193* or its chloride *194*. All cyclizations according to this procedure are accomplished using similar dilution principle parameters as already described for the acid chloride method, except that the leaving group of *196* replaces the assisting bases as a H^+-acceptor [225] and that the solvent dichloromethane [225] is used instead of benzene:

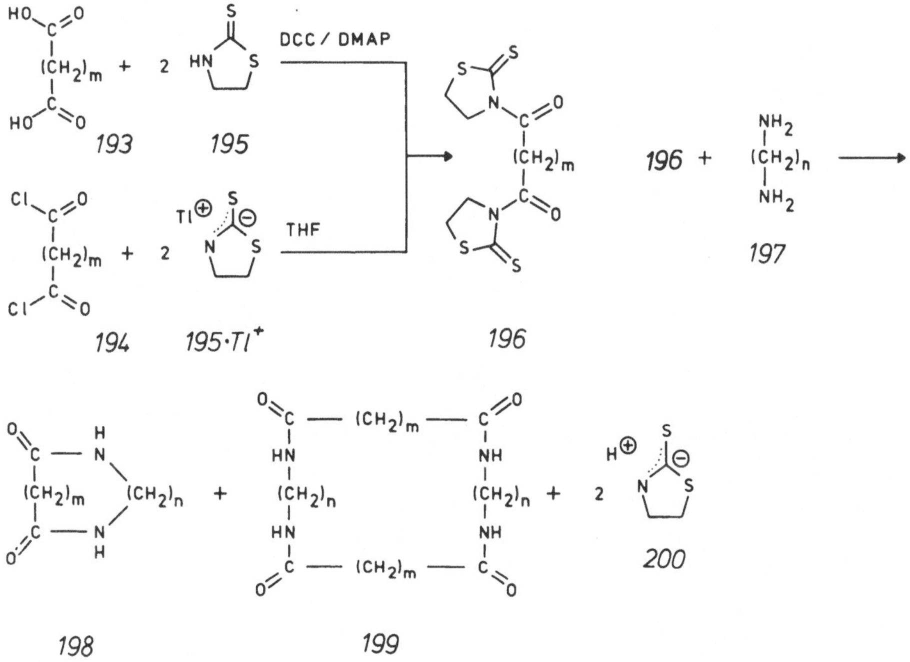

193 *195* *194* *195·Tl*$^+$ *196* *196* + *197* *198* *199* *200*

In the scope of the medium membered ring systems no monomeric diamides *198* were found, whereas with increasing ring size the diamides *198* (ring member number 14, 18, 22) as well as the dimeric tetraamides *199* (ring width 28, 36, 44) are formed [225]. These results are conformable to the frequently mentioned yield minimum [4].

In the following cyclization reaction not only the influence of *196* on the formation of monomeric and dimeric amides is of interest, but also its influence on the different reactivities of primary and secondary amines:

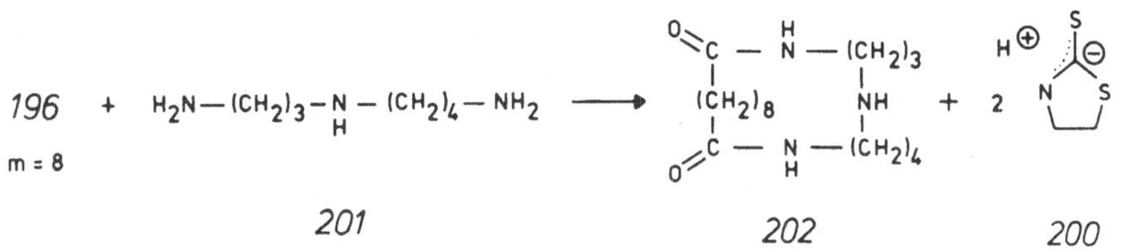

196 *201* *202* *200*

The two primary NH_2 groups in *201* react in 89 % yield under formation of the diamide *202* [225]. Byproducts, which are imaginable by C—N-bond formation with the secondary amino function, could not be indicated. This selectivity of primary amines using 3-acylthiazolidine-2-thiones is important for some natural product syntheses.

III.2.1.2 Crown Ether- and Cryptand Systems

As with the C—O-bond formation in Section II.2.1 yielding polyoxaethers, the substitution reactions at unsaturated C-atoms yielding azapolyethers can be divided into two large groups: the former includes medio- and macrocycles of such a kind which can be prepared by C—N-bond formation under dilution principle conditions. The second group includes ligand systems, the C—N-bond ring closure of which is achieved by the aid of the template effect [11, 13, 14] and/or the rigid group principle [11a, 12] without application of the dilution principle [1a].

As the thia- and polyoxaethers (cf. II.1.1.2, II.1.3.3, II.1.4.1, II.2.1) the first group also consists of molecular skeletons that contain the heteroatoms oxygen and sulfur besides nitrogen: More than that aromatic nuclei as well as heteroaromatic rings may be integrated into the cyclic ligands. Bi- and polycyclic systems, cryptands, can be constructed depending on the number and connexions of the nitrogen atoms.

Most of the known azapolyethers were synthesized according to the acid chloride method [222, 233–239, 241] applying the high dilution principle. Other possibilities for their synthesis [240, 241] using activated carboxyl components, as mentioned in section III.2.1 are of subordinated importance.

As the C—N-bond cyclizations yielding azapolyethers according to the Stetter cyclization [222] are analogous to the syntheses of aliphatic rings according to the same method (III.2.1.1). Details may be taken from the several reviews [11, 29, 109–111, 116, 119, 120], in which the syntheses and properties of nitrogen containing crown ethers are collected.

Here, a typical experimental procedure for the synthesis of the monocyclic crown ether amide *205* shall be given [233]:

| 203 | 204 | 205 |

starting components: a) 3,6-dioxaoctane-1,8-diamine (*203*) (99.8 mmole) in 500 ml of benzene
                      b) diacid dichloride *204* (44.4 mmole) in 500 ml of benzene
reaction type: 2C-DP [6a]
reaction medium: benzene (1.2 l)
reaction temperature: 10–20 °C
time of addition: 8 hrs
additional reaction time: none
yield: 75% of *205*

205 has been reduced also to give the corresponding amine in 75 % yield [233].

As described previously in the section III.2.1.1, minor variations of these dilution principle parameters were used for the syntheses according to the acid chloride procedure: a) high dilution cyclizations of this type are carried out as 2C-DP-reactions [6a]. b) The solvents to dissolve the educts and those in the reaction flask are generally benzene [84, 234, 235, 237–239, 241] or mixtures of benzene and trichloromethane [241] or DMF [236]. c) The reaction temperatures are far below the boiling temperatures of the solvents or their above mentioned mixtures. d) Either double molar amounts of the amine component or even higher excesses [233] are used as auxiliary bases, or a tertiary amine, e.g. triethylamine [237] is added to the reaction medium. e) The volume of the solvent and the time of addition depend on the amount of the used starting components. f) The yields vary strongly according to the structural type of the product formed.

III.2.1.3 Azaphanes

The synthesis of azaphane systems in accordance with the dilution principle is of theoretical interest; the influence of the reactivities of aromatic and aliphatic amines on the ring closure yield [242] and the hindered rotation of the C—N-bond in carboxylic acid amides [243a, c, 244] were investigated. A large number of the azaphane systems were synthesized under the aspect of peptide and enzyme modelling. Most of them were obtained according to the acid chloride procedure developed by Ruggli [5], Ziegler [4] and Stetter [222, 226, 242]. Nevertheless in some important dilution principle parameters the acid chloride method used for azaphanes [242] differs from the conditions of the Stetter cyclization [222] already described above in Sections III.2.1.1 and III.2.1.2. To characterize these differences, the conditions for the cyclizations of ortho- and para-phenylene diamines with aliphatic carboxylic acid dichlorides are compared [242]:

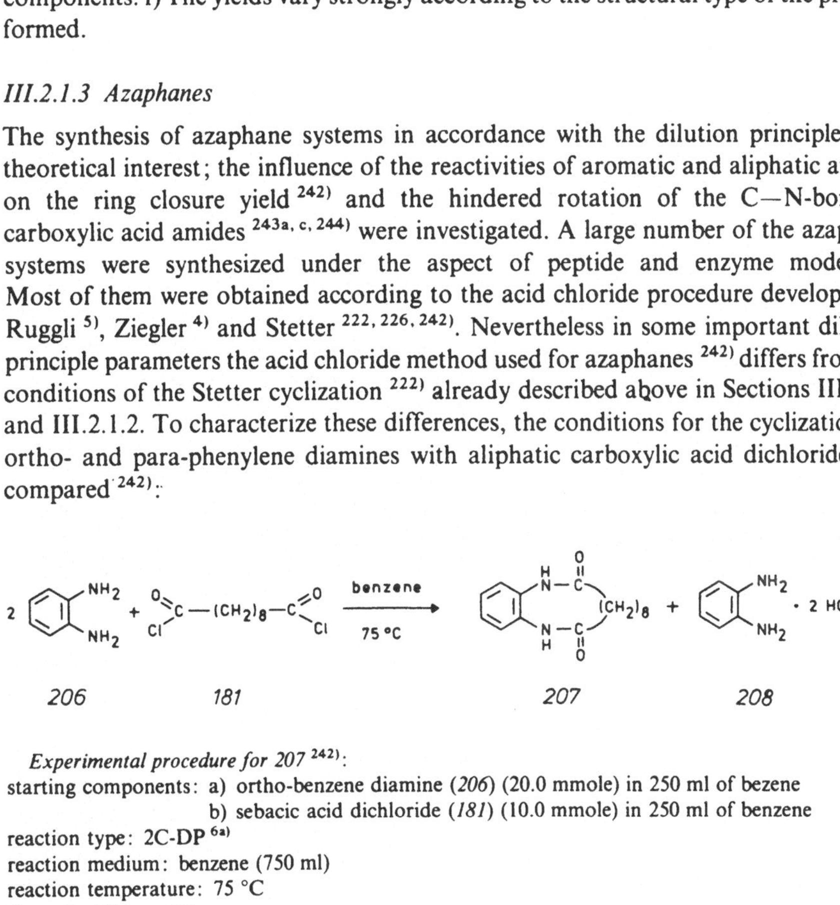

| 206 | 181 | 207 | 208 |

Experimental procedure for 207 [242]:

starting components: a) ortho-benzene diamine (206) (20.0 mmole) in 250 ml of bezene
 b) sebacic acid dichloride (181) (10.0 mmole) in 250 ml of benzene
reaction type: 2C-DP [6a]
reaction medium: benzene (750 ml)
reaction temperature: 75 °C
time of addition: 5 hrs
additional reaction time: 15 min
yield: 87.5% of 207

If pyridine is added as a HCl-acceptor, and equimolar amounts of 206 and 181 are used, the dilactam 207 is isolated in only 70% yield under otherwise similar conditions; no explanation for this result is given by the authors [242]. Due to the lower solubility of 210 in benzene the cyclization of the para-phenylene diamine (209) with sebacic acid dichloride (181) is carried out in dioxane at a molar ratio of 2:1. This solvent served for the dissolution of the reactants as well as a reaction medium.

$$209 \qquad 181 \qquad\qquad 210 \qquad\qquad 211$$

At a reaction temperature of 95 °C the para-cyclophane *210* is obtained in 35.4% yield [242]. The lower yield of *210* compared with *207* is explained by steric effects due to the hindered rotation around the C—N-bonds of the carbon·amide groups [242, 243a, c, 244].

These Stetter cyclizations [242] show the characteristic dilution parameters applied to the preparation of azaphanes: a) Most high dilution cyclizations leading to the amide functions are performed as 2C-DP-reactions [6a]. Only few reactions [243, b] follow the IC-DP-procedure [6a]. In these cases, the dissolved acid chloride is added dropwise to a solution of the amine and when occasion arises the auxiliary base. b) Unpolar solvents such as benzene [234a, b, 242, 245–249], toluene [251], or their mixtures [243a], orthodichlorobenzene [250], and also polar aprotic solvents [242, 252] and THF [243a, b] are used. c) Double molar amounts of the amines or auxiliary bases like pyridine [246] or triethylamine [243a] serve as HCl-acceptors. d) Many cyclizations are carried out in boiling solutions [242, 243b, 245, 246, 248, 250]. e) As already mentioned in the Sections III.2.1.1 and III.2.1.2 similar remarks are right for the addition times and the yields.

In the points b) and d) these dilution principle parameters for azaphane syntheses differ from the general conditions for the preparation of aliphatic ring systems (III.2.1.1) and azapolyethers (III.2.1.2). The behaviour of the solubility in a special solvent depends on the *diamine* used as starting component and the resulting diamide, as in the above mentioned case of the ortho- and para-phenylene diamines *206*, *209* and diamides *207*, *210* [242]. As the cyclizations usually lead to higher yields when carried out in homogenous solution, the optimal solvent is chosen with regard to the educts and products.

In consequence of the lower basicity of the *aromatic* amines [242] and of their slower reaction speed, such cyclizations are most often carried out in boiling solvents.

On the other hand, *benzylic* primary amines [249, 251] are most often cyclized at room temperature as their stability decreases with higher temperature. Benzylic amides are usually obtained in comparatively low yields despite of the increase of addition times and reaction times.

As a further example of an azaphane synthesized with regard to enzyme modelling, the reaction of N,N′-dimethyl-para-xylylene diamine (*212*), a secondary benzylic diamine with terephthalic acid dichloride (*161*) yielding the macrocyclic tetraamide *213* and the hexaamide *214* shall be described [248]:

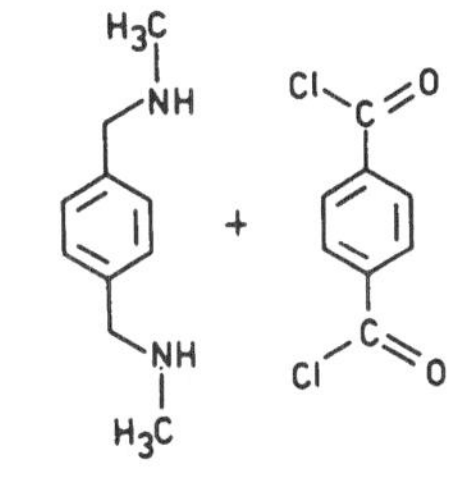

212   161

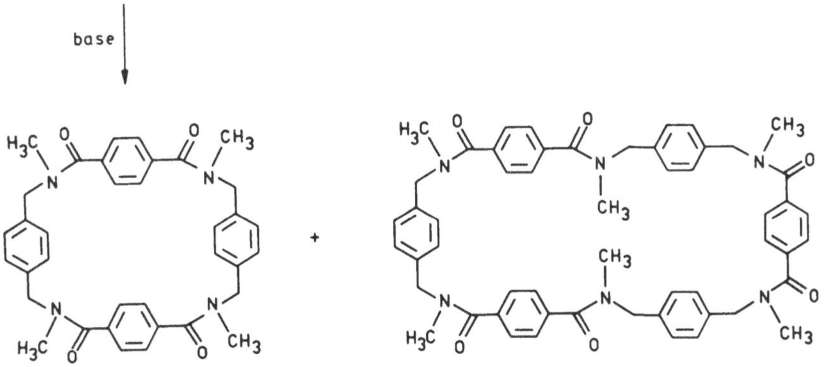

base

213   214

Under observance of the above described dilution principle conditions double molar amounts of *212* in boiling benzene lead to a total yield of 35% of *213* and *214* [248]. After hydrogenation of the amide bonds the corresponding tetra- and hexaamines are isolated in the ratio *213:214* = 1:6.75. The more rigid conformation of the tetra-amide *213* is made responsible for the lower yield: The four benzene rings are oriented face-to-face with each other, and they form a cavity with a diameter of approx. 4.5–5.5 Å [248]. The hexaamide *214* seems to be much more flexible and its aromatic rings are not in a fixed position; the maximal cavity diameter is estimated to be 8–10 Å [248].

For further amide cyclizations according to the acid chloride method yielding azaphanes we refer to the literature [242–252].

III.3 Synthesis of Medio- and Macrocyclic Compounds by Formation of C—C-Bonds

The Dieckmann condensation [253] and the Thorpe-Ziegler- (dinitrile-) reaction [253] belong to the nucleophilic substitution reactions at unsaturated C-atoms, which proceed under C—C-bond formation. The intermediates of these two cyclizations, a β-keto ester in the Dieckmann condensation and an enamino nitrile in the dinitrile reaction, are frequently not isolated, but directly hydrolysed and decarboxylated yielding the corresponding cyclic ketone. A disadvantage of these two reactions is that

61

medium membered ring ketones are only formed in low yields or even not at all. For this reason the acyloin condensation to be discussed below (cf. Sect. IV.1) is often favoured in this field.

III.3.1 C—C-Bond Formation by Dieckmann Condensation

Whereas the Dieckmann condensation constitutes a well suited procedure for the synthesis of small and normal carbocyclic compounds up to the present [254], it is rarely applied for the synthesis of medium- and many-membered rings nowadays for reasons mentioned above (III.3) [253]. Intra- and intermolecular ring closure reactions yielding aliphatic medio- and macrocycles, therefore are mainly described in older studies [255-262]. Cyclophane syntheses [263-265] via ester condensation have only become known in very few cases. Not only because of the low number of phane systems, but also because of the similar cyclization conditions for aromatic and aliphatic rings no division into cycloaliphatic systems and phanes is necessary here.

Beginning with diesters, carbocycles and heterocyclic compounds containing N-, O-, S-heteroatoms [257-262], were synthesized (see below).

The ester cyclization method worked out by Leonard [255, 257-260, 262] seems to be generally applicable, the characteristic dilution parameters shall be put on record here:

Experimental procedure for the cyclization of the diesters 215a,b of differing chain length to yield the 10- and 15-membered ketones 218a,b respectively their dimers 219a,b [255]:

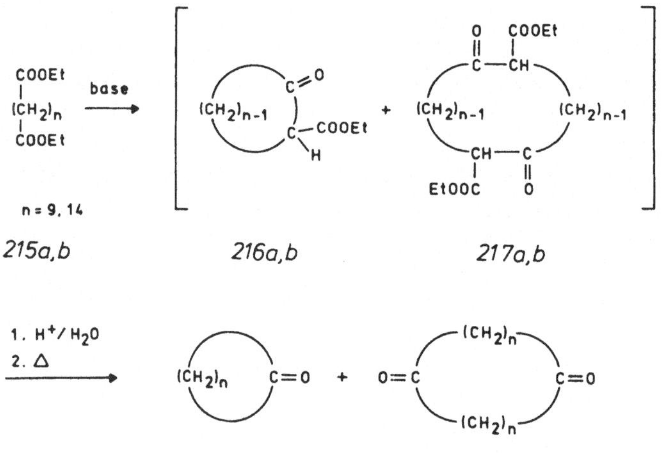

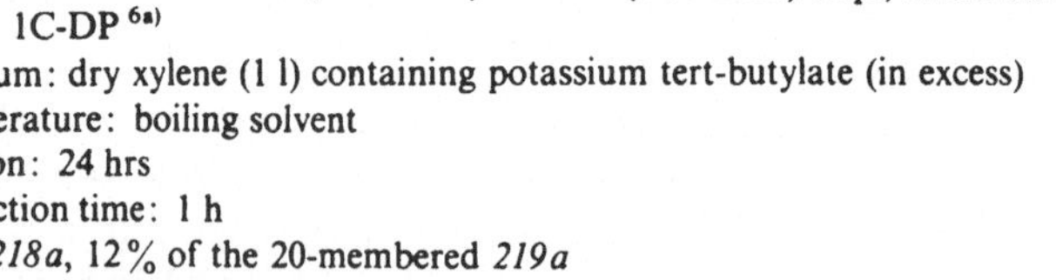

starting component: diester *215a* (0.05 mole) or *215b* (0.05 mole) resp., in 250 ml of xylene
reaction type: 1C-DP [6a]
reaction medium: dry xylene (1 l) containing potassium tert-butylate (in excess)
reaction temperature: boiling solvent
time of addition: 24 hrs
additional reaction time: 1 h
yield: 0% of *218a*, 12% of the 20-membered *219a*
  The result is different in the *b*-series:
yield: 48% of *218b*, 0.9% of the 30-membered diketone *219b*

General features of the high dilution Dieckmann condensation are: a) they all are carried out as 1C-DP-reactions [6a], b) unpolar solvents like xylene are used for the reaction components and as a reaction medium, c) the ring closure reaction takes place in boiling solvents, d) alkali metal hydroxides, hydrides and alcoholates of sodium and potassium as well as potassium tert-butylate are employed as bases for the formation of the carbanion, they are often applied in large excess, e) the dropping times and the reaction times are usually long; they range from a minimum of 24 hrs [255] to a maximum of 9 days [256], f) the yields of β-keto esters and of ketones respectively are very low in the 9–12-membered ring series. The yields increase with growing ring size. Higher yields exceeding 50%, are not attainable as a rule.

In the heterocyclic series, only 8-membered rings and the corresponding dimeric products were particularly investigated [257–262], with the aim to get insight into trans-anular interactions between heteroatoms and the carbonyl group through the space:

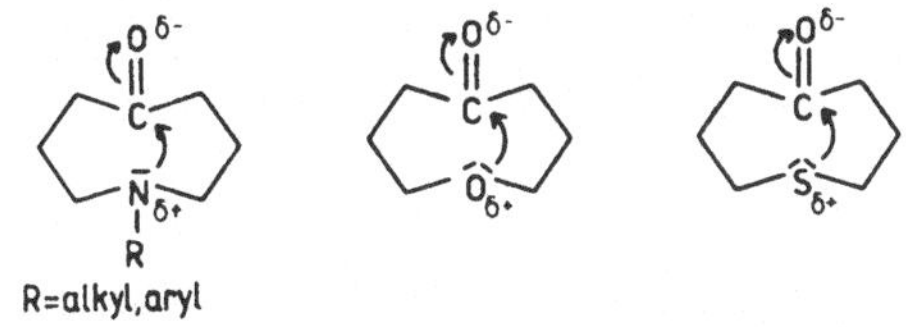

R=alkyl,aryl

220   221   222

The use of the strong bases such as mentioned in point d), is disadvantageous as only such educts can be used which do not contain base sensitive functions. This problem was solved in the ring closure reaction yielding zearalanone (223) [265] (cf. III.1.1.3) which contains a base sensitive lactone function:

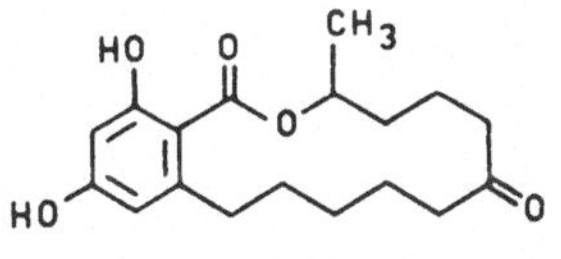

223

The voluminous base sodium-bis(trimethylsilyl)amide, $NaN[Si(CH_3)_3]_2$ [165], was used. Its basicity is high enough to enter into a reaction with CH-acidic esters under formation of a sodium enolate and bis(trimethylsilyl)amine. The base is also completely dissolved in organic solvents like ether, so that the dilution reaction can be carried out in homogeneous solution. After successful preliminary tests with aliphatic diesters [265] in ether this base was also employed for the cyclization of the benzyl protected ester 224:

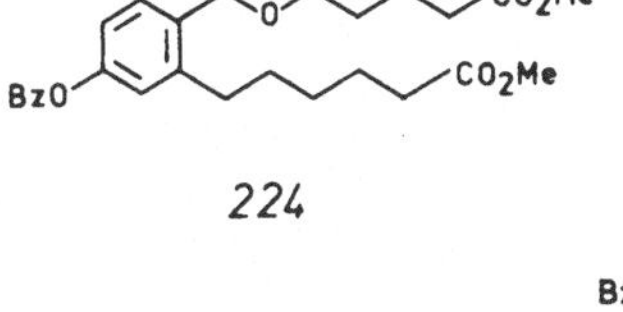

224

$NaN[Si(CH_3)_3]_2$

225a   +   225b

63

Experimental procedure for 225 a,b [265)]:
starting component: triester *224* (3.00 mmole) in 260 ml of dry ether
reaction type: 1C-DP [6a)]
reaction medium: ether (175 ml) containing NaN[Si(CH$_3$)$_3$]$_2$ (19.0 mmole)
reaction temperature: boiling solvent
time of addition: 8 hrs
additional reaction time: 15 min
yield: 77% of *225a* and *225b*

A separation of these isomers wasn't tackled as both of them form the same ketone after hydrolysis and decarboxylation.

For further ester condensation reactions according to the dilution principle we refer to the original literature [254–265)].

III.3.2 C—C-Bond Formation using Dinitriles (Thorpe-Ziegler Reaction)

This reaction was developed by Thorpe [268)] as a synthetic method for 5–7-membered rings [268a)] and for the chain lengthening of open chained compounds [268b)]. Later, the reaction was extended to medium- and many-membered ring systems by Ziegler [9, 87, 269–272)].

The Thorpe-Ziegler reaction [266, 267)] shall be compared here with the Dieckmann condensation [253, 254)] (Sect. III.3.1). After their hydrolysis and decarboxylation both reactions yield cyclic ketones — often without isolation of the cyclic intermediates. Both methods show marked yield minima [4)] in the region of the medium-membered ring systems [4, 266)]. Therefore other C—C-coupling methods (cf. Sects. III.4.1, III.4.2 and IV.1) are preferred in this area today. As modern studies are lacking in this field and as aliphatic and aromatic ring systems were synthesized using similar dilution conditions, the above used classification into cycloalkanes [270–277)] and phanes [269, 278, 279)] can be skipped here. A difference between the Thorpe-Ziegler reaction and the Dieckmann condensation is due to the dinitrile as the starting component. The cyclic intermediate is formulated as an enamino nitrile today [266, 267, 280)]. The following dilution conditions seem to be general valid for Thorpe-Ziegler reactions, as can be drawn from Ziegler's cyclizations of dicyano compounds yielding aliphatic ring systems [87, 270–272)] as well as cyclophanes [269)].

Experimental procedure for the dilution principle cyclization yielding 5-tert-butylcyclooctanone (228) [273)]:

226   227   228

starting component: 4-tert-butyl-1,7-dicyanoheptane (*226*) (100 mmole) in 900 ml of ether
reaction type: 1C-DP [6a)]
reaction medium: ether (1.5 l) containing lithium N-methylanilide
reaction temperature: boiling solvent
time of addition: 48 hrs
additional reaction time: no instructions given
yield: 77% of *228*, without isolation of the intermediate enamino nitrile *227*

The general conditions may be summarized as follows:

a) All dinitrile cyclizations are carried out as 1C-DP-reactions [6a]. b) Dry ether as a solvent for the base and for the dinitrile is most often used [266]. Only in ring closure reactions yielding cyclophanes [278] sometimes ether/benzene mixtures are used to dissolve the educt. c) All dinitrile ring closures are accomplished at the boiling temperature of ether. d) Sodium and lithium salts of secondary amines exclusively come into question. Alkyl anilides as condensation reagents are favoured. The preparation of the amides is performed in the reaction flask directly before the cyclization reaction under inert conditions. e) The addition times vary; usually they amount to 24 hrs. f) The yields for 8-, 14- and higher membered rings are remarkable high for phanes as well as for cycloaliphatic rings; as a rule they exceed 50%.

Compared with the typical dilution principle parameters of a diester cyclization [254] differences are to be noted as regards the points b), d) and f).

To b) above: Although the educts of both reaction types are cyclized in boiling solvents, the dinitrile condensations proceed under milder conditions due to the large boiling point differences between ether and xylene (34.6° and approx. 140 °C).

To d) above: In the Dieckmann condensation [254] the optimal base must be discovered experimentally, whereas in the dinitrile cyclizations the alkyl anilides can be regarded as the best condensation reagents.

To f) above: The Thorpe-Ziegler reaction [266] is superior regarding the yields of larger ring systems, whereas they are usually lower than 50% in the diester cyclization [254]. Also, lower amounts of byproducts are usually obtained.

Despite these advantages the dinitrile cyclization was not widely used in the last years. This may to a certain extent be due to the more general applicability of the Dieckmann intermediate β-ketoester, which can further be functionalized in various ways.

III.4 Synthesis of Medio- and Macrocyclic Compounds by Formation of C=C-Double-Bonds

The Wittig reaction and one of its variations, the phosphonate method, can be accomplished under dilution principle conditions to yield intra- or intermolecular ring closure products [281]. For simplification we treat this reaction type under topic III.

III.4.1 C=C-Double Bond Formation by Wittig Reaction

Intramolecular C=C-bond formations via Wittig reaction [282] were rather rarely described; in this case details of the dilution conditions often are not evident.

On the other hand, numerous intermolecular ring closures via bis-Wittig reactions [282] with precise data of the experimental procedure have been reported. In these ring closure reactions mainly aromatic components are introduced. Therefore often phanes are the products [282]. An example for a bis-Wittig cyclization reaction is the following [282]:

x^1, x^2 = aromatic or aliphatic systems

229   230   231

There are two general ways *A* and *B* to carry out a cyclizing double Wittig reaction [282]:

A) The bis-ylid *230* is generated from the bis-phosphonium salt using a suitable base prior to the cyclization step. The ring closure of the bis-carbonyl component *229* takes place under dilution principle conditions. Procedure *A* therefore always is a 1C-DP-reaction [6a]. The low stability of many bis-ylids of the type *230* limitates this procedure [282].

In these cases the second method *B* may be applied: Here, the bis-ylids *230* are generated in situ from the corresponding phosphonium salts by bases. The experimental procedure can follow two ways: Either the base is added dropwise to equivalent amounts of the dialdehyde *229* and the bis-phosphonium salt (1C-DP-reaction [6a]), or the ring closure reaction yielding *231* is performed as a 2C-DP-reaction [6a]. In the latter case, the educts *229* and the bis-phosphonium salt from the one and the base from the other dropping funnel are simultaneously added into a certain solvent volume in the reaction flask. The alternative procedure *B* is limited by the low stability of many aldehydes in basic media [282].

Both methods shall be illuminated by focussing attention on their experimental characteristics:

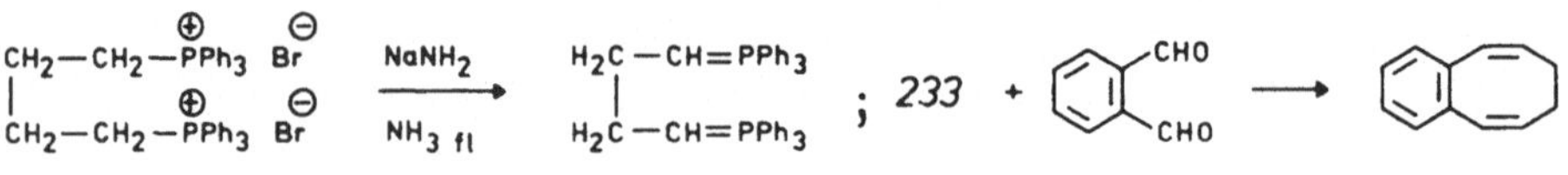

| *232* | *233* | *234* | *235* |

Experimental procedure for 1,2-benzocycloocta-1,3,7-triene (235) [283]:

starting component: ortho-phthalic dialdehyde (*234*) (60.0 mmole) in 300 ml of ether
reaction type: 1C-DP [6a]
reaction medium: ether (300 ml) containing phosphonium ylid *233* produced from tetramethylene-bis-triphenylphosphonium dibromide (*232*) (60.0 mmole) and sodium amide (135.0 mmole)
reaction temperature: room temperature
time of addition: 30 min
additional reaction time: 30 min, afterwards 20 hrs at 70 °C
yield: 18% of *235*

The bases used for the preparation of the ylids are the following [282]: a) organo lithium compounds in ether, hexane or benzene; b) sodium amide in liquid ammonia; b) alcoholates in alcohols and/or DMF; d) dimethylsulfoxylate in DMSO [282].

The great selection of bases and solvents results in an important dilution principle condition, in which both methods *A* and *B* differ. In method *A* the solvents of the educt solutions usually correspond to the reaction medium in which the phosphonium ylid was generated: they may be polar/protic, polar/aprotic and unpolar.

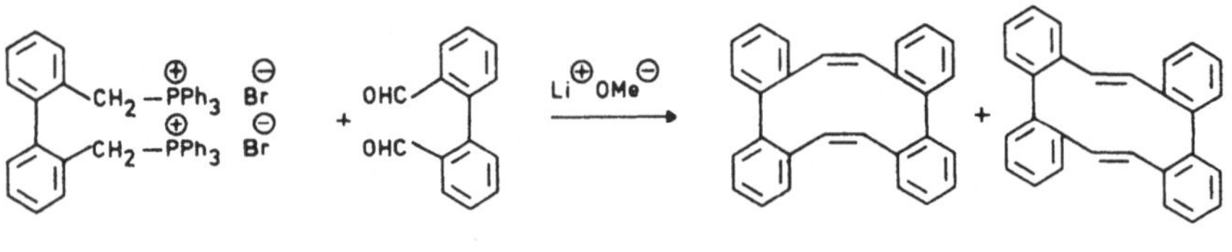

| *236* | *237* | *238a* | *238b* |

Experimental procedure for the isomers 238a and 238b according to method B [282, 284]:
starting components: a) 2,2'-biphenyl dialdehyde (*237*) (24.3 mmole) and
2,2'-bis(triphenylphosphoniomethyl)biphenyl dibromide (*236*) (24.3 mmole) in 500 ml of methanol
b) lithium (0.1 gat) in 500 ml of methanol
reaction type: 2C-DP [6a]
reaction medium: dry methanol (4 l)
reaction temperature: room temperature
time of addition: 12.5 hrs
additional reaction time: 12 hrs
yield: 1.1% of *238a* and 4.2% of *238b*

The same reaction was also carried out according to method A, but it only yielded traces of the 12-membered phane systems *238* [285].

Method B is different from method A in the following three parameters: a) the cyclization may be accomplished as 1C- or as 2C-DP-reactions [6a] as mentioned above. In these reactions the base for the formation of the ylid may not be put forward into the reaction flask, but must be added dropwise as one component independent of the other educts. b) As a rule, only lithium alkoxylates are used as bases. c) This limits the variation of the solvents. Usually, the alcohol corresponding to the alcoholate or its mixture with DMF is chosen.

The following dilution conditions are used for both methods:
a) the reaction temperatures vary. Partly the intermolecular ring closure reactions proceed in boiling solvents, partly only after the addition of the component(s) the reaction mixture is heated, and other bis-Wittig reactions [282] are carried out at room temperature. b) The addition and reaction times vary strongly. The dosage on an average takes several hours. The total reaction time as a rule is about one day. c) The yields differ strongly. Depending on the type of the product they range from 0.03% (traces) to more than 50%.

The mechanistic factors which influence the stereochemistry of the C=C-double bonds to be formed were already discussed frequently [286–288]. The configuration of the C=C-double bond depends on the type of the ylid, the carbonyl compound, the solvent and on the presence or absence of salts [282].

III.4.2 C=C-Double Bond Formation with Phosphonates

A newer method for the synthesis of ring systems containing C=C-double bonds is the phosphonate method [189, 289–291]. Phosphonate anions show several advantages as against the phosphorane ylids of the Wittig reaction [281] described in the literature [281, 292].

Whereas the configuration of the alkenes, obtained by the usual Wittig reaction, depends on the phosphorane ylids and carbonyl compounds used as starting components and of the reaction conditions [286–288], these factors play a subordinate role in the formation of Z- and E-alkenes by the phosphonate method. In most cases, the E-configurational isomer is the only product; stereoselectivity is an important advantage of the phosphonate method [289]. The cyclization using phosphonates seems to be the method of choice for the synthesis of cyclic compounds of all types and all ring width [189, 293] and of natural products [189, 290, 294, 295] today, such as (—)-vermiculin, carbonolide B and muscone for example. A typical phosphonate reaction according to the dilution principle shall be noted here.

Experimental procedure for the 13- and 15-membered α,β-unsaturated lactones 240a and 240b [293]:

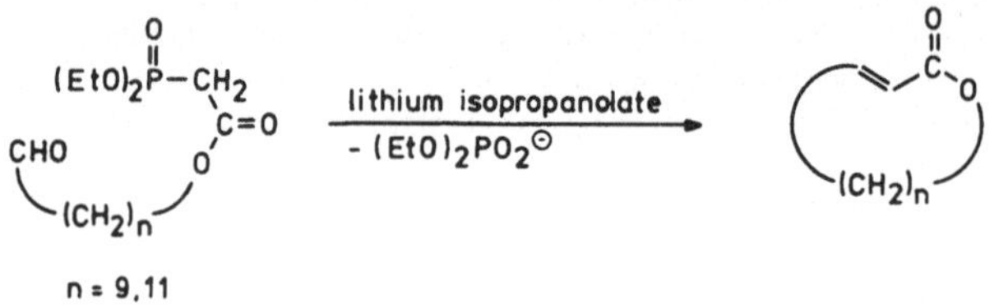

n = 9,11

239a,b   *240a,b*

starting components: a) phosphonate *239a* or *239b* resp. (0.50 mmole) in 5 ml of THF/benzene (1:1)
                  b) lithium isopropanolate (0.55 mmole) in 5 ml of THF
reaction type: 2C-DP [6a]
reaction medium: THF (50 ml) and HMPA (0.5 ml)
reaction temperature: 20 °C
time of addition: 15 hrs
additional reaction time: 1 h
yield: 60–70% and 40–50% of *240a* and *240b* resp.

Exclusively the *E*-configurated products are isolated.

The formation of dimeric dilactones by intermolecular ring closure is negligibly low; their yields lie under 1%. As with most dilution principle reactions a yield decrease occurs with increasing concentrations of the starting components. The general features of a phosphonate reaction according to the dilution principle can be summarized as follows:

a) The nucleophilic phosphonate carbanions are always generated in situ. In contrast to the usual Wittig reaction the base, necessary for the proton abstraction, can be submitted in the reaction flask or may simultaneously be dropped to the reaction mixture. Consequently, inter- and intramolecular cyclizations according to the phosphonate method can be carried out as 1C- or as 2C-DP-reactions [6a]. b) Different base types are suitable for the formation of the phosphonate anion. Most often, lithium and sodium alkoxylates and sodium hydride are used. The lithium hexamethyl-disilazane is rarely required. c) The solvents may either be polar, aprotic or unpolar (THF, HMPA, DME, benzene); however, no solvents with protic properties are used. d) The reaction temperature is important. Most often room temperature is favoured in connection with the phosphonate method. e) The yields around 60% on an average are rather good, compared with the phosphorane Wittig reaction.

As an example for the synthesis of a cyclic natural product we choose the 15-membered d,l-muscone (*241*) [294]:

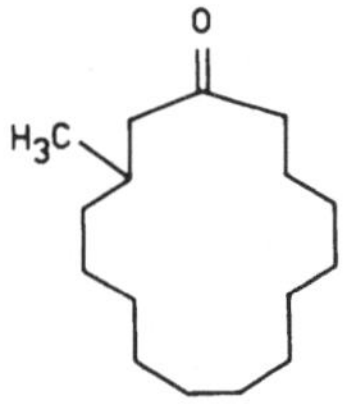

241

A short and efficient synthetic method for this cycloalkanone is the phosphonate procedure, which is superior to the Dieckmann condensation (III.3.1) and the Thorpe-Ziegler cyclization (III.3.2).

Beginning with Z-oleic acid (*242*), the ketophosphonate derivative *243* is obtained in few steps:

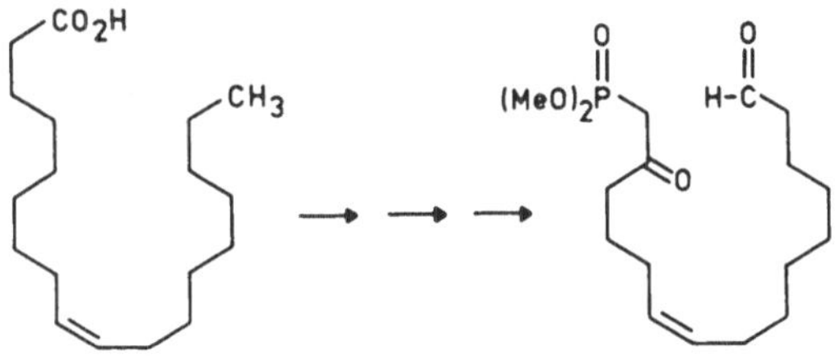

The cyclization of *243* with sodium hydride in DME yields the carbocyclic compound as a mixture of Z- and E-isomers *244* in a total yield of 50%. The separation of both configurational isomers is not necessary, as, after methylation with lithium dimethylcuprate and hydrogenation with Pd/C, both of them lead to the desired d,l-muscone (*241*) in 90% yield. The cyclic diketone *245* is also obtained in a noticeable yield of 15–20% by an intermolecular ring closure.

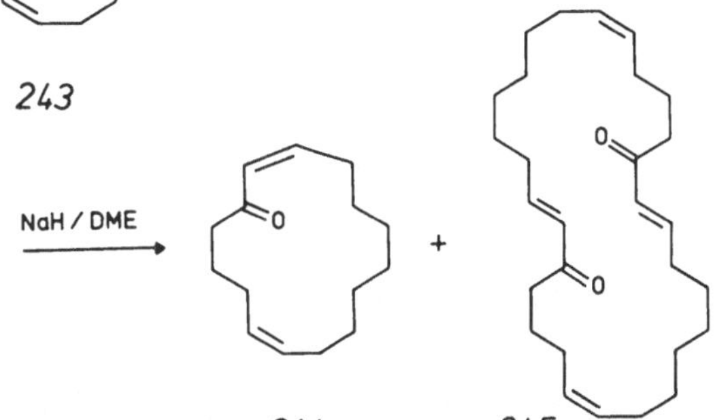

Further informations regarding the phosphonate cyclization method is given in the literature [189, 281, 282, 289 – 291, 293 – 295].

IV Reduction Reactions for the Synthesis of Medio- and Macrocyclic Compounds

IV.1 C—C-Bond Formation by Acyloin Condensation

The acyloin condensation has stood the test for the preparation of medium- and many-membered ring systems according to the dilution principle. The diester, acting as the starting component, is reduced with metals to yield a cyclic acyloin, an α-hydroxy ketone, by C—C-coupling.

Acyloins often only serve as intermediate products [296, 297] for the preparation of cyclic derivatives: By catalytic reduction 1,2-diols (glycols) are available which can be transferred into alkenes, e.g. by the use of the Corey-Winter method [298]. The Clemmensen reaction is quoted for the selective reduction of the hydroxy group of an acyloin yielding the cyclic ketone. Stronger reaction conditions lead to the reduction of both the OH- and carbonyl function yielding the hydrocarbon. Acyloins can also be oxidized to 1,2-diketones which are starting components for the synthesis of other derivatives, e.g. cyclic acetylenes.

This versatile applicability of the α-hydroxy ketones, exceeding the preparation of medio- and macrocyclic acyloins, has often been used, e.g. for the preparation of perfume components, e.g. muscone (241) [297], for the synthesis of heterocycles [296] and cyclophanes [297]. The acyloin condensation often is the only cyclization method which leads, in high yields, to the desired ring system, especially in the region of the medium-membered cycles where other C—C-coupling methods (Dieckmann condensation, III.3.1 or dinitrile reaction III.3.2) often fail.

The ring closure always takes place under similar dilution principle conditions, independent of the additional functional groups of the starting diester as e.g. unsaturated or aromatic unities or heteroatoms like nitrogen, oxygen and sulfur. Therefore a subdivision of the hitherto studied acyloin condensation reactions into cycloalkanes [299–307] and phanes is superfluous [168, 308–310]. Numerous phane systems were synthesized by using this method [168, 296, 297, 308–310].

For an example we point to a more general applicable dilution principle cyclization yielding sebacoin, a 10-membered aliphatic acyloin that was described in detail by Allinger [299] in Organic Synthesis; sebacoin is obtained in a yield of 63 to 66 % [299].

We compare the ring closure reactions starting from the Z- and E-alkene diester 246 yielding the 10-membered cyclic compounds Z- and E-247 [303, 305].

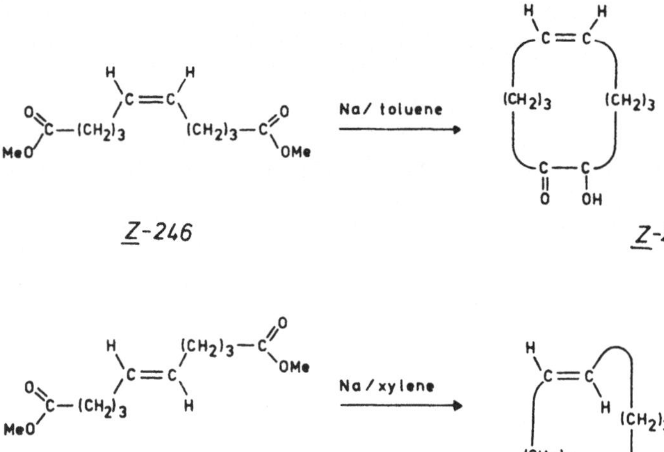

$$Z\text{-}246 \quad \xrightarrow{\text{Na/ toluene}} \quad Z\text{-}247$$

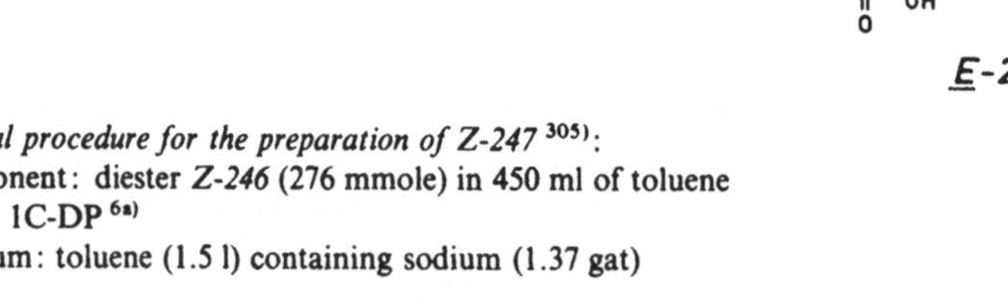

$$E\text{-}246 \quad \xrightarrow{\text{Na/ xylene}} \quad E\text{-}247$$

Experimental procedure for the preparation of Z-247 [305]:
starting component: diester Z-246 (276 mmole) in 450 ml of toluene
reaction type: IC-DP [6a]
reaction medium: toluene (1.5 l) containing sodium (1.37 gat)

reaction temperature: boiling solvent
time of addition: 18 hrs
additional reaction time: 1 hr
yield: 78% of Z-247

The ring closure giving E-247 was carried out under otherwise similar dilution principle conditions in boiling xylene and led to a yield of 51% [303].

Characteristic features of dilution principle reactions of the acyloin ring closure method shall be summarized here:

a) All acyloin condensations are performed as 1C-DP-reactions [6a]. b) The reactions must be carried out under an inert gas atmosphere [296, 297, 311]. c) The reducing metal is used at least in a double molar amount per ester function; for a diester therefore 4 equivalents are necessary. Most often sodium [168, 299–310], only in a few cases potassium [313] was employed. Sodium/potassium alloys [312, 313] offer an advantage in effecting reducing at low temperatures which is useful for natural product synthesis. d) Unpolar aprotic solvents for educts as well as for the reaction solvent such as toluene and xylene [168, 299–312] are applied. Mixtures of these two solvents are common too. Polar aprotic solvents like ether [313] or dioxane [314] more rarely are applied. e) The reaction temperature usually corresponds to the boiling temperature of the solvent or solvent mixture. f) The addition times vary strongly. They usually take several hrs. g) The yields are rather high for all ring types and ring widths [297].

The higher yield of the Z-product 247 (78%) [305] compared with E-247 (51% yield) [303] corresponds to the expectations [297]. The Z-configuration contributes to bring the terminal ester functions in a position, which is favourable for the ring closure step and the Z-product 247 is energetically favoured. This is due to the lower transanular interactions of the Z-oriented H-atoms. In E-configurated compounds these interactions are sometimes high enough to lead to the corresponding hydrogenated cycles in appreciable yields. For summing up: E-double bonds as rigid groups [3, 12] in 10-membered rings do not increase the yields. A remarkable fact in this connection is that no mediocyclic acyloins containing an acetylenic bond [297, 303, 305] in 5-position could be indicated. If the chain length of a linear acetylene diester is long enough yet, the ring closure step is facilitated by putting down the steric interactions of the H-atoms in the interior of the ring [305].

Similarly, the synthesis of paracyclophanes is limited by the chain length of the bridge. The shortest bridge, constructed by the aid of the acyloin condensation, contains 9 CH_2 groups ([9]paracyclophane).

Especially in aliphatic cyclic compounds with medium ring members the alkoxylate anions, set free in the course of the reduction, can evoke side reactions [296, 312, 315–317]. As strong bases these anions are able to split off the H-atom α-positioned to the ester function and to react according to a Dieckmann condensation (III.3.1). This competing reaction is preferred if the ring system, thereby formed, has an energetical more favourable conformation than the cycles, formed by the acyloin condensation. In the acyloin cyclization the homologous ring containing one additional CH_2-group is formed [311].

This side reaction can be largely eliminated if the acyloin condensation is carried out in the presence of trimethylchlorosilane (TMCS) [315, 316]. TMCS is applied in equimolar amounts related to the reducing agent. The efficiency of TMCS to catch alkoxylates results from equations 1–3 [316]:

$$(1) \quad (CH_2)_x \begin{array}{c} CO_2R \\ CO_2R \end{array} \xrightarrow{+4\,Na} (CH_2)_x \begin{array}{c} C-O^{\ominus}\,Na^{\oplus} \\ \| \\ C-O^{\ominus}\,Na^{\oplus} \end{array} + 2\,RO^{\ominus}\,Na^{\oplus}$$

$$(2) \quad (CH_2)_x \begin{array}{c} C-O^{\ominus}\,Na^{\oplus} \\ \| \\ C-O^{\ominus}\,Na^{\oplus} \end{array} + 2\,TMCS \longrightarrow (CH_2)_x \begin{array}{c} C-OTMS \\ \| \\ C-OTMS \end{array} + 2\,NaCl$$

$$(3) \quad 2\,RO^{\ominus}\,Na^{\oplus} + 2\,TMCS \longrightarrow 2\,ROTMS + 2\,NaCl$$

TMCS itself forms silyl endiol ethers (cf. equation 2) with the sodium salts of the endiolates (cf. equation 1), which can be transferred into the acyloin by an acidic hydrolysis. The Na$^+$ ions thereby are bound as NaCl. On the other hand, TMCS reacts with the sodium salts of the alkoxylates yielding silyl ethers under formation of NaCl (cf. equation 3). The medium remains neutral during the reaction in this way.

The use of TMCS has proved to be successful in many acyloin condensations especially to avoid side reactions such as Dieckmann cyclizations and β-eliminations [312, 315–317].

V Oxidation Reactions for the Synthesis of Medio- and Macrocyclic Compounds

V.1 C—C-Bond Formation by Oxidative Coupling of Acetylenes

Several variations [318–320] for the oxidative coupling of terminal diacetylenes are accepted. Only one method yet is suitable for the synthesis of medio- and macrocyclic systems according to the dilution principle. This method, named after its inventors, Eglinton and Galbraith [320, 321], consists in the oxidation of terminal acetylenes with copper salts as an oxidant. The dilution principle reactions are carried out in a non-watery, homogeneous and weakly basic solution. The range of application of the Eglinton/Galbraith method [321] is broad. Unsaturated and saturated, aliphatic and aromatic hydrocarbon cycles as well as ester and ketone ring systems were prepared by the aid of this method, including natural products [193, 290, 321, 322] as e.g. the 16-membered cyclic lactone exaltolide (153) (cf. Sect. III.1.1.1).

The products of the C—C-bond formation by a cyclizing oxidation of two acetylenic groups contain four sp-hybridised C-atoms neighboured to each other [323], forming an angle of 180°. In the mediocyclic range (8–10 ring members) high geometrical strains are effective besides the Pitzer strain and transanular interactions, due to the deformation of the 180°-angle. This directs to a low thermodynamic stability, and therefore 8- to 11-membered rings, if at all are only formed in situ or at low temperatures [323]. With increasing ring members the stability increases likewise. The yields quickly increase with increasing ring width in like manner. This may be due to the C—C-triple bond acting as a rigid group [3, 11a, 12]. It lowers the conformational mobility of the alkane chain and favours the ring closure.

Cycloaliphatic rings [322, 324–326] and phane systems [327–332] need the same dilution principle parameters which we describe below selecting one member of both compound classes:

248   *249*

Experimental procedure for the annulenone 249 [326]:
starting component: diacetylene *248* (1.50 mmole) in 34 ml of pyridine/ether 3:1
reaction type: 1C-DP [6a]
reaction medium: pyridine/ether 3:1 (99 ml) containing dry cupric acetate (8.49 mmole)
time of addition: 2 hrs
reaction temperature: 50–55 °C
additional reaction time: 30 min
yield: 48 % of *249*

This procedure exhibits the characteristic dilution conditions for aliphatic and for aromatic acetylene ring closure reactions, which can be summarized as follows:
a) All oxidative acetylene couplings according to the Eglinton/Galbraith [321] procedure are carried out as 1C-DP-reactions [6a]. b) The oxidizing reagent — cupric acetate [321, 323] — is used in high excess, up to the 8-fold amount related to the acetylenic educt. c) The educt component is generally dissolved in pyridine [321, 323] to neutralize the generated acetic acid. In contrast to the acidic reaction medium a weak basic medium does not disturb the cyclization procedure. In many cases also combinations of pyridine with ether and/or methanol are applied, whereby the amount of pyridine is always preponderantly [326]. d) The reaction temperatures vary only slightly. Most often ring closure reactions are carried out in a boiling solvent, sometimes at only slightly elevated temperatures. e) Also the addition times differ only little. On an average they total around 3–4 hrs. f) The yields vary strongly. They range from a minimum of 5.5 % to a maximum of 88 %. It is difficult to derive rules for the dependence of the yields on factors like ring width, functional groups, inter- and intramolecular reaction or rigid groups [3, 11a, 12] and so on.

From a stereochemical view the [4.4]orthocyclophane-1,3,11,13-tetraine (*253*) [327–329] is interesting. The X-ray analysis shows that the 12-membered ring *253* is nearly planar and that there is an angle between the triple bonds [321]. Moreover two centers of high π-electron density face each other leading to transanular interactions [321, 323]. Based on the unexpected low stability of *253* it was presumed at first [327, 328] that the cyclization of ortho-diethinylbenzene (*250*) would lead to the strain-free trimeric product *251*, but already before the result of the X-ray analysis was given out, this presumption could be disproved [321, 328].

73

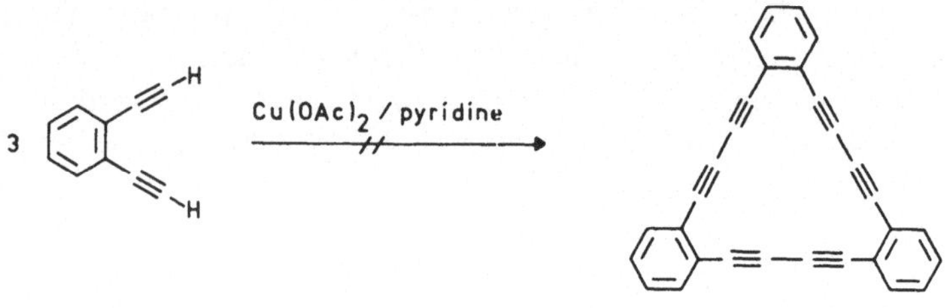

250   251

The dimeric phane 253 was synthesized according to the dilution principle proceeding from two different educts by an inter- as well as an intramolecular ring closure step [327−329]:

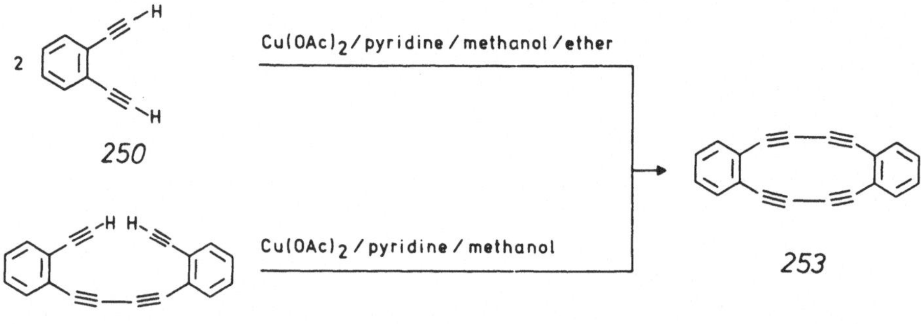

250

252   253

Under the above mentioned dilution conditions ortho-diethinylbenzene (250) in pyridine/methanol/ether (6:6:1) was changed into the phane 253 in 43% yield[323] by an intermolecular ring closure reaction. A negligible yield increase to 46% [323] is obtained in the intramolecular reaction beginning with 252 under analogous dilution principle conditions without addition of ether. The 12-membered cycle 253 is only stable at low temperatures in benzene or pyridine, and it decomposes under explosion if heated or powdered. After some time the yellow crystals change their colour to black even under exclusion of air and light. These properties underline the supposed low stability of the phane 253 [323].

At the end a synthesis of a natural product shall be described to indicate the broad application area mentioned at the beginning of the section regarding the oxidative C—C-coupling: the synthesis of the macrolide exaltolide (153) [193, 196a, 290, 321, 322]:

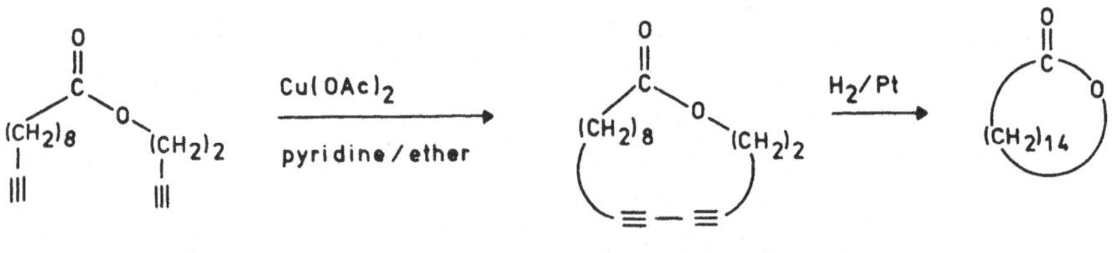

254   255   153

Exaltolide is a saturated 16-membered lactone ring, the synthesis of which begins with the acyclic terminal diacetylene *254*. As described above the educt *254* is cyclized to give the lactone ring *255*. The high yield of *255* (88%) [322] is remarkable. The catalytic hydrogenation directs to the desired natural product exaltolide (*153*) (1,15-pentadecanolide), which has the characteristic agreeable smell of the oil isolated from the plant Angelica archangelica officinalis [322].

Some ring systems are known, the preparation of which was carried out according to the conventional Glaser coupling [318] reaction without applying the dilution principle. This method consists in the C—C-coupling of diterminal acetylenes with Cu_2Cl_2, watery ammonium chloride solution and oxygen as an oxidizing reagent. As these reactions are performed in inhomogenous phase, they are less suited for the preparation of cyclic systems. Yet, the Glaser coupling was applied by Sondheimer [333-335] for the synthesis of a row of theoretically interesting annulenes.

VI Closing Remarks

This survey shall facilitate the planning of new medio- and macrocyclic ring syntheses via using the dilution principle by the collection and comparison of the known facts, of approved methods and procedures in twofold respect: On the one hand, essential points and suggestions shall be extracted for the selection of the suitable reaction type and on the other hand for the optimization of the experimental conditions. Both factors together possibly allow a prediction concerning the approximate yields to be expected for their trends if other dilution conditions are used.

The classical reaction type leading to many ring systems is *a priori* predetermined: A polyoxaether as a rule will be synthesized by C—O-bond formation (II.2.1) and a thiaphane by C—S-bond coupling (II.1.1.3, II.1.2, II.1.3.2, II.1.4.2). Nevertheless, some ring systems can be synthesized on different reaction paths: This, for example, is valid for medium- and many-membered ring ketones. Here, several methods can be used, e.g. the Dieckmann condensation (III.3.1), the dinitrile reaction (III.3.2) and the acyloin condensation (IV.1). In the corresponding sections the advantages and disadvantages of these methods are discussed comparatively so that a decision can be taken for or against a specific reaction type.

A detailed experimental procedure for a specific high dilution cyclization is of interest, too. There was no space to include detailed experimental descriptions, but the reaction schemes and the listing of some of the reaction parameters should allow comparisons and realizations of the efforts, to be made, how long the reaction takes to be accomplished and which type of reagent must be used. These short descriptions of the procedures serve as a quick information and orientation of the employed concentrations and volumes, temperatures and apparatus [6-10, 23, 24]. General valid dilution parameters for each C—X-bond coupling reaction were drawn up.

From the data given, the inter- or intramolecular course of a reaction may be predicted approximately and the yields of the possible oligomers may be estimated to some extent.

It would be valuable and useful to discuss the above dilution principle reaction types in the light of the following parameters in a somewhat more quantitative manner: degree of dilution [1-4], reaction speeds, E. M. (effective molarity [1a]), influence of solvent [1-4], caesium effect [13, 14], gauche-effect [11a], apparatus used [6-10, 23, 24] and so on. At the "state of the art" available at present, this seems not to be possible in a straightforward way for most of the reactions described. For senseful generalisations and conclusions often systematic studies are lacking as a presumption. Here broad physico-chemical studies are missing.

In the last years new methods have been developed to synthesize medio- and macrocyclic systems beginning with acyclic educts [15a-j]. Template [11] and caesium effects [13, 14] were introduced and the gauche-effect [11a] was discussed. Also methods using tin [11g-i] and palladium [15i, j] organic compounds to preshape cycles open new paths to many-membered rings [336].

This survey may facilitate the decision which way can or cannot be gone in a distinct synthetic question.

VII Acknowledgements

We are grateful to Dr. E. Weber (Bonn) and Dr. P. Neumann (Ludwigshafen) for their advice. Dipl.-Chem. M. Wittek and Dipl.-Chem. M. Palmer helped with the drawings, Mrs. R. Mahr with the translation. Thanks are due to Mrs. B. Jendrny for the preparation of the manuscript. Support from the "Fonds der Chemischen Industrie" and from the science ministery of NRW is gratefully acknowledged.

VIII References

1. a) Galli, C., Mandolini, L.: J. C. S. Chem. Commun. *1982*, 251
   b) Galli, C, Illuminati, G., Mandolini, L.: J. Org. Chem. *45*, 311 (1980)
   c) Dalla-Cort, A., Mandolini, L., Masci, B.: ibid. *45*, 3923 (1980)
   d) Review: Illuminati, G., Mandolini, L.: Acc. Chem. Res. *14*, 95 (1981)
   e) Illuminati, G., Mandolini, L., Masci, B.: J. Am. Chem. Soc. *103*, 4142 (1981)
   f) Review: Winnik, M. A.: Chem. Rev. *81*, 491 (1981)
2. a) Adams, R., Whitehill, L. N.: J. Am. Chem. Soc. *63*, 2073 (1941)
   b) Salomon, G.: Helv. Chim. Acta *17*, 851 (1934)
   c) Salomon, G.: ibid. *19*, 1079 (1936)
   d) Ruzicka, L., Salomon, G., Meyer, K. E.: ibid. *20*, 109 (1937)
   e) Stoll, M., Rouvé, A.: ibid. *17*, 1284 (1934)
   f) Reichardt, C.: Lösungsmitteleffekte in der organischen Chemie. Verlag Chemie. Weinheim/ Bergstraße, 1968; 2. Aufl. 1973. Reichardt, C.: Solvent Effects in Organic Chemistry. Verlag Chemie, Weinheim 1979
   g) Dalla-Cort, A., Illuminati, G., Mandolini, L., Masci, B.: J. Chem. Soc. Perkin Trans. II, *1980*, 1774
3. Cf.: Neumann, P.: Dissertation Univ. Heidelberg, 1973, pp. 51
4. Review: Ziegler, K. in: Methoden der Organischen Chemie (Houben-Weyl-Müller), 4. Aufl., Bd. IV/2, pp. 73, G. Thieme, Stuttgart, 1955
5. Ruggli, P.: Liebigs Ann. Chem. *392*, 92 (1912)
6. a) Vögtle, F.: Chem.-Ztg. *96*, 396 (1972)
   b) Vögtle, F.: Chem. Ind. (London) *1973*, 1037
7. Böckmann, K.: Dissertation Univ. Bonn, 1980, p. 18
8. Karbach, S., Löhr, W., Vögtle, F.: J. Chem. Res. (S) *1981*, 314; (M) *1981*, 3579
9. a) Hammerschmidt, E., Vögtle, F.: ibid. *1980*, 192; (M) *1980*, 2776
   b) Hammerschmidt, E., Schlütter, H., Vögtle, F.: ibid. *1980*, 86; (M) *1980*, 1083
10. Balaam, D. F., Chippindall, J., Davy, J. R., Reiss, J. A.: Chem. Ind. (London) *1975*, 354

11. a) Review: Laidler, D. A., Stoddart, J. F.: "Synthesis of crown ethers and analogues" (Patai, S. ed.), The Chemistry of functional groups, Suppl. E., part 1, John Wiley and .Sons, Chichester–New York–Brisbane–Toronto 1980, p. 1–15
    b) De Sousa Healy, M., Rest, A. J.: Adv. Inorg. Chem. Radiochem. *21*, 1 (1978)
    c) Bowsher, B. R., Rest, A. J.: J. C. S. Dalton *1981*, 1157
    d) Greene, R. N.: Tetrahedron Lett. *1972*, 1793
    e) Cook, F. L., Caruso, T. C., Byrne, M. P., Bowers, C. W., Speck, D. H., Liotta, C. L.: ibid. *1974*, 4092
    f) Kulstad, S., Malmsten, L. Å.: Tetrahedron *36*, 521 (1980)
    g) Shanzer, A., Mayer-Sochet, N.: J. C. S. Chem. Commun. *1980*, 176
    h) Shanzer, A., Berman, E.: ibid. *1980*, 259
    i) Steliou, K., Szczygielska-Nowosielska, A., Favre, A., Poupart, M. A., Hanessian, S.: J. Am. Chem. Soc. *102*, 7578 (1980)
12. a) Baker, W., McOmie, J. F. W., Ollis, W. D.: J. Chem. Soc. *1951*, 200
    b) Baker, W., Clark, D., Ollis, W. D., Zealley, T. S.: ibid. *1952*, 1452
    c) Drewes, S. E., Coleman, P. C.: J. C. S. Perkin I *1972*, 2148
    d) Drewes, S. E., Riphagen, B. G.: ibid. *1974*, 323 and 1908
    e) Drewes, S. E., Colemann, P. C.: ibid. *1974*, 2578
    f) Thulin, B., Wennerström, O., Somfai, I., Chmielarz, B.: Acta Chem. Scand. *B31*, 135 (1977)
13. a) Kruizinga, W. H., Kellogg, R. M.: J. Am. Chem. Soc. *103*, 5183 (1981)
    b) Buter, J., Kellogg, R. M.: J. C. S. Chem. Commun. *1980*, 466
    c) Buter, J., Kellogg, R. M.: J. Org. Chem. *46*, 4481 (1981)
14. a) Vögtle, F., Klieser, B.: Synthesis *1982*, 249
    b) Klieser, B., Vögtle, F.: Angew. Chem. *94*, 632 (1982); Angew. Chem., Int. Ed. Engl. *21*, 618 (1982); Angew. Chem. Suppl. *1982*, 1392–1397
    c) Weber, E., Vögtle, F., Josel, H.-P., Newkome, G. R., Pucket, W. E.: Chem. Ber. *1983*, in press
15. a) Hopf, H., Lenich, F. Th.: Chem. Ber. *107*, 1891 (1974)
    b) Gilb, W., Menke, K., Hopf, H.: Angew. Chem. *89*, 177 (1977); Angew. Chem., Int. Ed. Engl. *16*, 191 (1977)
    c) Böhm, I., Herrmann, H., Menke, K., Hopf, H.: Chem. Ber. *111*, 523 (1978)
    d) Newkome, G. R., Majestic, V. K., Fronczek, F. R.: Tetrahedron Lett. *1981*, 3035
    e) Newkome, G. R., Majestic, V. K., Fronczek, F. R.: ibid. *1981*, 3039
    f) Nakashita, Y., Hesse, M.: Angew. Chem. *93*, 1077 (1981); Angew. Chem., Int. Ed. Engl. *20*, 1021 (1981)
    g) Kramer, U., Guggisberg, A., Hesse, M., Schmid, H.: Angew. Chem. *90*, 210 (1978); Angew. Chem., Int. Ed. Engl. *17*, 200 (1978)
    h) Stork, G., Macdonald, T. L.: J. Am. Chem. Soc. *97*, 1264 (1975)
    i) Trost, B. M.: Pure and Appl. Chem. *59*, 787 (1979)
    j) Trost, B. M., Verhoeven, T. R.: J. Am. Chem. Soc. *102*, 4743 (1980)
    k) Masamune, S., Kamada, S., Schilling, W.: J. Am. Chem. Soc. *97*, 3515 (1975)
16. a) Jäger, V., Buß, V., Schwab, W.: Liebigs Ann. Chem. *1980*, 122. It should be mentioned here that even compounds of normal ring width were synthesized advantageously according to the dilution principle, e.g. some oxazolines (cf. also reference [9b])
    b) Serratosa, F., Lopez, F., Font, J.: An. Quim. *70*, 893 (1974)
17. a) von Braun, J.: Ber. Dtsch. Chem. Ges. *43*, 3220 (1910)
    b) Grischewitsch-Trochimowski, E.: Chem. Zentralbl. *1923* III, 773
    c) Shriner, R. L., Struck, H. C., Jorison, W. J.: J. Amer. Chem. Soc. *52*, 2066 (1930)
    d) Bost, R. W., Conn, M. W.; Org. Synth., Coll. Vol. II, 547 (1943)
18. a) Müller, A., Schütz, A. F.: Ber. Dtsch. Chem. Ges. *71*, 692 (1938)
    b) Müller, A., Funder-Fritzsche, E., Konar, W., Rintersbacher-Wlasak, E.: Monatsh. Chem. *84*, 1206 (1953)
19. Friedman, P., Allen, Jr., P.: J. Org. Chem. *30*, 780 (1965)
20. Friedman, P., Allen, Jr., P.: ibid. *27*, 1095 (1962)
21. Mandolini, L., Vontor, T.: Synth. Commun. *9*, 857 (1979)
22. a) Hunsdiecker, H., Erlbach, H.: Chem. Ber. *80*, 132 (1947)
    b) Schöberl, A., Gräfje, H.: Liebigs Ann. Chem. *614*, 66 (1958)

23. Vögtle, F.: Chem. Ind. (London) *1972*, 346
24. Harvard Apparatus Corp., Dover, Mass., Model 600, -000
25. Review: Cope, A. C., Martin, M. M., McKervey. M. A.: Chem. Rev., Chem. Soc. *20*, 119 (1966)
26. Allinger, N. L., Zalkov, V.: J. Org. Chem. *25*, 701 (1960)
27. Prelog, V.: J. Chem. Soc. *1950*, 420
28. a) Borgen, G., Gaupset, G.: Acta Chem. Scand. *B 28*, 816 (1974)
    b) Borgen, G., Dale, J.: ibid. *26*, 159 (1972)
29. Review: a) Vögtle, F., Weber, W.: Kontakte (Merck) *1/77*, 11; b) ibid. *2/77*, 16; c) ibid. *3/77*, 36; d) ibid. *2/78*, 16; e) Vögtle, F., Weber, E., Elben, U.: ibid. *2/80*, 36
30. Bradshaw, J. S., Hui, J. Y., Haymore, B. L., Christensen, J. J., Izatt, R. M.: J. Heterocycl. Chem. *10*, 1 (1973)
31. Nomenclature of polyether systems: Pedersen, C. J., Frensdorff, H. K.: Angew. Chem. *84*, 16 (1972); Angew. Chem., Int. Ed. Engl. *11*, 16 (1972). For nomenclature of thiapolyethers cf. ref. [30]
32. a) Dann, J. R., Chiesa, P. P., Gates, Jr., J. W.: J. Org. Chem. *26*, 1991 (1961)
    b) Review: Bradshaw, J. S., Hui, J. Y., Chan, Y., Haymore, B. L., Izatt, R. M., Christensen, J. J.: J. Heterocycl. Chem. *11*, 45 (1974)
33. Review: a) Vögtle, F., Neumann, P.: Fortschr. Chem. Forsch. *48*, 67 (1974)
    b) Vögtle, F., Neumann, P.: Synthesis *1973*, 85
    c) Vögtle, F., Neumann, P.: Chimia *26*, 64 (1972)
    d) Review: Vögtle, F., Hohner, G.: Top. Curr. Chem. *74*, 1 (1978)
    e) Vögtle, F., Neumann, P.: Angew. Chem. *84*, 75 (1972); Angew. Chem., Int. Ed. Engl. *11*, 73 (1972)
34. a) Vögtle, F., Förster, H.: Angew. Chem. *89*, 443 (1977); Angew. Chem., Int. Ed. Engl. *16*, 429 (1977)
    b) Vögtle, F., Neumann, P.: Tetrahedron *26*, 5299 (1970)
35. Stevens, T. S.: Prog. Org. Chem. *7*, 48 (1968)
36. a) Bruhin, J., Jenny, W.: Tetrahedron Lett. *1973*, 1215
    b) Boekelheide, V., Reingold, I. D., Tuttle, M.: J.C.S., Chem. Commun. *1973*, 406
    c) Rebafka, W., Staab, H. A.: Angew. Chem. *85*, 831 (1973); Angew. Chem., Int. Ed. Engl. *12*, 776 (1973)
    d) Bieber, W., Vögtle, F.: Chem. Ber. *111*, 1653 (1978)
37. Review: Vögtle, F., Rossa, L.: Angew. Chem. *91*, 534 (1979); Angew. Chem., Int. Ed. Engl. *18*, 515 (1979)
38. Vögtle, F.: Angew. Chem. *81*, 258 (1969); Angew. Chem., Int. Ed. Engl. *8*, 274 (1969)
39. For nomenclature of phanes cf.:
    a) Smith, B. H.: Bridged Aromatic Compounds, Chapter 1, Academic Press, New York—London 1964
    b) Vögtle, F., Neumann, P.: Tetrahedron *26*, 5847 (1970)
40. a) Vögtle, F., Schunder, L.: Liebigs Ann. Chem. *721*, 129 (1969)
    b) Vögtle, F.: Chem. Ber. *102*, 1449 (1969)
41. Vögtle, F., Schunder, L.: Chem. Ber. *102*, 2677 (1969)
42. Martel, H. J. J.-B., Rasmussen, M.: Tetrahedron Lett. *1971*, 3843
43. Mitchell, R. H., Boekelheide, V.: J. Am. Chem. Soc. *92*, 3510 (1970)
44. Boekelheide, V., Anderson, P. H.: J. Org. Chem. *38*, 3928 (1973)
45. Sato, T., Wakabayashi, M., Kainosho, M., Hata, K.: Tetrahedron Lett. *1968*, 4185
46. Vögtle, F., Schäfer, R., Schunder, L., Neumann, P.: Liebigs Ann. Chem. *734*, 102 (1970)
47. Miyahara, Y., Inazu, T., Yoshino, T.: Chem. Lett. *1980*, 397
48. Vögtle, F., Lichtenthaler, R. G.: Synthesis *1972*, 480
49. Lichtenthaler, R. G., Vögtle, F.: Chem. Ber. *106*, 1319 (1973)
50. Sato, T., Wakabayashi, M., Hata, K., Kainosho, M.: Tetrahedron *27*, 2737 (1971)
51. In ref. [50] there are also described the syntheses of phanes containing one or more *disulfide bridges*
52. Reviews:
    a) Griffin, Jr., R. W.: Chem. Rev. *63*, 45 (1963)
    b) Sato, T., Akabori, S., Kainosho, M., Hata, K.: Bull. Chem. Soc. Jpn. *39*, 856 (1966)
53. Mitchell, R. H., Boekelheide, V.: Tetrahedron Lett. *1970*, 1197

54. Boekelheide, V., Hollins, R. A.: J. Am. Chem. Soc. *95*, 3201 (1973)
55. Davy, J. R., Reiss, J. A.: Tetrahedron Lett. *1972*, 3639
56. Vögtle, F.: Chem.-Ztg. *95*, 668 (1971)
57. Bieber, W., Vögtle, F.: Angew. Chem. *89*, 199 (1977); Angew. Chem., Int. Ed. Engl. *16*, 175 (1977)
58. Hammerschmidt, E., Bieber, W., Vögtle, F.: Chem. Ber. *111*, 2445 (1978)
59. Vögtle, F., Hammerschmidt, E.: Angew. Chem. *90*, 293 (1978); Angew. Chem., Int. Ed. Engl. *17*, 268 (1978)
60. Wingen, R., Vögtle, F.: Chem. Ber. *113*, 676 (1980)
61. Hammerschmidt, E., Vögtle, F.: ibid. *113*, 1121 (1980)
62. Hammerschmidt, E., Vögtle, F.: ibid. *113*, 3550 (1980)
63. a) Vögtle, F., Grütze, J.: Angew. Chem. *87*, 543 (1975); Angew. Chem., Int. Ed. Engl. *14*, 559 (1975)
    b) Grütze, J., Vögtle, F.: Chem. Ber. *110*, 1978 (1977)
64. Vögtle, F., Atzmüller, M., Wehner, W., Grütze, J.: Angew. Chem. *89*, 338 (1977); Angew. Chem., Int. Ed. Engl. *16*, 325 (1977)
65. a) Böckmann, K., Vögtle, F.: Chem. Ber. *114*, 1048 (1981)
    b) Böckmann, K., Vögtle, F.: ibid. *114*, 1065 (1981)
66. Bradshaw, J. S., Hui, J. Y.: J. Heterocycl. Chem. *11*, 649 (1974)
67. Review: Reid, E. E.: "Organic Chemistry of Bivalent Sulfur", Vol. III, Chemical Publ. Co., Inc., New York, 1960, pp. 12–42
68. Stetter, H., Wirth, W.: Liebigs Ann. Chem. *631*, 144 (1960)
69. Vögtle, F., Effler, A. H.: Chem. Ber. *102*, 3071 (1969)
70. Boekelheide, V., Anderson, P. H., Hylton, T. A.: J. Am. Chem. Soc. *96*, 1558 (1974)
71. Akabori, S., Shiomi, K., Sato, T.: Bull. Chem. Soc. Jpn. *44*, 1346 (1971)
72. a) Vögtle, F., Lichtenthaler, R.: Z. Naturforsch. *26b*, 872 (1971)
    b) Vögtle, F., Neumann, P., Zuber, M.: Chem. Ber. *105*, 2955 (1972)
    c) Vögtle, F., Zuber, M., Neumann, P.: Z. Naturforsch. *26b*, 707 (1971)
    d) Vögtle, F., Grütze, J., Nätscher, R., Wieder, W., Weber, E., Grün, R.: Chem. Ber. *108*, 1694 (1975)
    e) Weber, E., Wieder, W., Vögtle, F.: ibid. *109*, 1002 (1976)
    f) Vögtle, F.: Chem.-Ztg. *94*, 313 (1970)
    g) Vögtle, F.: Liebigs Ann. Chem. *735*, 193 (1970)
    h) Vögtle, F.: Chem. Ber. *102*, 3077 (1969)
73. a) Sherrod, S. A., da Costa, R. L., Barnes, R. A., Boekelheide, V.: J. Am. Chem. Soc. *96*, 1565 (1974)
    b) Lawson, J., DuVernet, R., Boekelheide, V.: ibid. *95*, 956 (1973)
    c) Gundermann, K.-D., Röker, K.-D.: Angew. Chem. *85*, 451 (1973); Angew. Chem., Int. Ed. Engl. *12*, 425 (1973)
    d) Kawashima, T., Otsubo, T., Sakata, Y., Misumi, S.: Tetrahedron Lett. *1978*, 5115
    e) Kannen, N., Umemoto, T., Otsubo, T., Misumi, S.: ibid. *1973*, 4537
    f) Nakazaki, M., Yamamoto, K., Toya, T.: J. Org. Chem. *45*, 2553 (1980)
74. Vögtle, F., Wolz, U.: Chem. Exp. Didakt. *1*, 15 (1975)
75. Mitchell, R. H., Boekelheide, V.: J. Am. Chem. Soc. *96*, 1547 (1974)
76. Staab, H. A., Haenel, M.: Chem. Ber. *106*, 2190, 2203 (1973)
77. Haenel, M., Staab, H. A.: Tetrahedron Lett. *1970*, 3585
78. Umemoto, T., Otsubo, T., Misumi, S.: ibid. *1974*, 1573
79. a) Bottino, F., Pappalardo, S.: Tetrahedron *36*, 3095 (1980)
    b) Bottino, F., Foti, S., Pappalardo, S.: ibid. *33*, 337 (1977)
    c) Cf. Fehér, F., Glinka, K., Malcharek, F.: Angew. Chem. *83*, 439 (1971); Angew. Chem., Int. Ed. Engl. *10*, 413 (1971)
    The references [79a–c] deal with the formations of di- and trisulfide compounds according to the thiolate method; cf. reference [50]
    d) Bottino, F., Foti, S., Pappalardo, S.: Tetrahedron *32*, 2567 (1976)
80. Staab, H. A., Herz, C. P., Döhling, A.: Chem. Ber. *113*, 233 (1980)
81. Ochrymowycz, L. A., Mak, C.-P., Michna, J. D.: J. Org. Chem. *39*, 2079 (1974)
82. a) Black, D. St. C., McLean, I. A.: Tetrahedron Lett. *1969*, 3961
    b) Black, D. St. C., McLean, I. A.: Chem. Commun. *1968*, 1004

83. a) Travis, K., Busch, D. H.: ibid. *1970*, 1041
    b) Rosen, W., Busch, D. H.: J. Am. Chem. Soc. *91*, 4694 (1969)
    c) Rosen, W., Busch, D. H.: Inorg. Chem. *9*, 262 (1970)
    d) Linday, L. F., Busch, D. H.: J. Am. Chem. Soc. *91*, 4690 (1969)
84. a) Weber, E., Vögtle, F.: Liebigs Ann. Chem. *1976*, 891
    b) Cf. Gould, E. S.: "Mechanismus und Struktur in der organischen Chemie", 2nd ed., pp. 674, Verlag Chemie, Weinheim 1969
85. Meadow, J. R., Reid, E. E.: J. Am. Chem. Soc. *56*, 2177 (1934)
86. Review: Sicher, J.: Progr. Stereochem. *3*, 215 (1962)
87. Ziegler, K., Holl, H.: Liebigs Ann. Chem. *528*, 143 (1937)
88. Vögtle, F., Weber, E.: Angew. Chem. *86*, 126 (1974); Angew. Chem., Int. Ed. Engl. *13*, 149 (1974)
89. Dale, J.: J. Chem. Soc. III, *1963*, 93
90. Otsubo, T., Stusche, D., Boekelheide, V.: J. Org. Chem. *43*, 3466 (1978)
91. Kamp, D., Boekelheide, V.: ibid. *43*, 3470 (1978)
92. Boekelheide, V., Galuszko, K., Szeto, K. S.: J. Am. Chem. Soc. *96*, 1578 (1974)
93. a) Davy, J. R., Iskander, M. N., Reiss, J. A.: Aust. J. Chem. *32*, 1067 (1979)
    b) Davy, J. R., Reiss, J. A.: ibid. *29*, 163 (1976)
    c) Jessup, P. J., Reiss, J. A.: ibid. *29*, 173 (1976)
    d) Boekelheide, V., Tsai, C.-H.: Tetrahedron *32*, 423 (1976)
    e) Leach, D. N., Reiss, J. A.: Tetrahedron Lett. *1979*, 4501
    f) Longone, D. T., Küsefoglu, S. H., Gladysz, J. A.: J. Org. Chem. *42*, 2787 (1977)
    g) Davy, J. R., Iskander, M. N., Reiss, J. A.: Tetrahedron Lett. *1978*, 4085
    h) Haenel, M. W.: ibid. *1977*, 4191
    i) Otsubo, T., Misumi, S.: Synthetic Commun. *8*, 285 (1978)
    j) Reingold, J. D., Schmidt, W., Boekelheide, V.: J. Am. Chem. Soc. *101*, 2121 (1979)
    k) Gray, R., Boekelheide, V.: ibid. *101*, 2128 (1979)
    l) Gray, R., Boekelheide, V.: Angew. Chem. *87*, 138 (1975); Angew. Chem., Int. Ed. Engl. *14*, 107 (1975)
94. Vögtle, F., Wester, N.: Liebigs Ann. Chem. *1978*, 545
95. Vögtle, F., Steinhagen, G.: Chem. Ber. *111*, 205 (1978)
96. a) Rebafka, W., Staab, H. A.: Angew. Chem. *85*, 831 (1973); Angew. Chem., Int. Ed. Engl. *12*, 776 (1973)
    b) Diederich, F., Staab, H. A.: Angew. Chem. *90*, 383 (1978); Angew. Chem., Int. Ed. Engl. *17*, 372 (1978)
97. Haenel, M. W., Flatow, A., Taglieber, V., Staab, H. A.: Tetrahedron Lett. *1977*, 1733
98. Rossa, L., Vögtle, F.: J. Chem. Res. (S) *1977*, 264; (M) *1977*, 3010
99. Atzmüller, M., Vögtle, F.: Chem. Ber. *111*, 2547 (1978)
100. a) Hammerschmidt, E., Dissertation Univ. Bonn 1980
     b) Hammerschmidt, E., Vögtle, F.: Chem. Ber. *112*, 1785 (1979)
101. Hohner, G., Vögtle, F.: ibid. *110*, 3052 (1977)
102. Vögtle, F., Hohner, G., Weber, E.: J. C. S. Chem. Commun. *1973*, 366
103. Vögtle, F., Winkel, J.: Tetrahedron Lett. *1979*, 1561
104. Hendrickson, J. B., Cram, D. J., Hammond, G. S. in: Organic Chemistry, 3rd ed., (Hugus, Z. Z., Margerum, D. W., Richards, J. H., Yankwich, P. E., eds.), McGraw-Hill, Kogakushu, LTD., 1970, pp. 393
105. Pedersen, C. J.: J. Am. Chem. Soc. *89*, 2495, 7017 (1967)
106. a) Lüttringhaus, A.: Liebigs Ann. Chem. *528*, 181 (1937)
     b) Lüttringhaus, A.: ibid. *528*, 211 (1937)
     c) Lüttringhaus, A.: ibid. *528*, 223 (1937)
107. Lüttringhaus, A., Ziegler, K.: ibid. *528*, 155 (1937)
108. Review: Vögtle, F., Weber, E.: "Crown Ethers — Complexes and Selectivity" (Patai, S. ed.), The Chemistry of Functional Groups, Suppl. E, part 1, John Wiley and Sons, Chichester — New York — Brisbane — Toronto, 1980, p. 59
109. Review: a) Dalley, N. K.: in: "Synthetic Multidentate Macrocyclic Compounds", (Izatt, R. M., Christensen, J. J., eds.), Academic Press, New York — San Francisco — London, 1973, Chap. 4, p. 207-243

b) Hiraoka, M.: "Crown Compounds — their characteristics and applications", Elsevier Scientific Publ. Co. Amsterdam—Oxford—New York, 1982

110. Review: Lehn, J.-M.: Struct. Bonding *16*, 1 (1973)
111. Review: Izatt, R. M., Christensen, J. J.: Progress in Macrocyclic Chemistry, Wiley, New York, 1979
112. Graf, E., Lehn, J.-M.: J. Am. Chem. Soc. *98*, 6403 (1976)
113. a) Tabushi, I., Gasaki, H., Kuroda, Y.: ibid. *98*, 5727 (1976)
     b) Tabushi, I., Kuroda, Y., Kimura, Y.: Tetrahedron Lett. *1976*, 3327
114. a) Review: Vögtle, F., Sieger, H., Müller, W. M.: Top. Curr. Chem. *98*, 107 (1981)
     b) Vögtle, F., Müller, W. M., Weber, E.: Chem. Ber. *113*, 1130 (1980)
     c) Weber, E., Vögtle, F.: Angew. Chem. *92*, 1067 (1980); Angew. Chem., Int. Ed. Engl. *19*, 1030 (1980)
115. a) Dehmlow, E. V.: Angew. Chem. *86*, 187 (1974); Angew. Chem., Int. Ed. Engl. *13*, 170 (1974)
     b) Dehmlow, E. V.: Angew. Chem. *89*, 521 (1977); Angew. Chem., Int. Ed. Engl. *16*, 493 (1977)
116. Review: Christensen, J. J., Eatough, D. J., Izatt, R. M.: Chem. Rev. *74*, 351 (1974)
117. Newkome, G. R., Sauer, J. D., Roper, J. M., Hager, D. C.: ibid. *77*, 513 (1977)
118. Reviews: a) Bradshaw, J. S. in: "Synthetic Multidentate Macrocyclic Compounds", (Izatt, R. M., Christensen, J. J., eds.), Academic Press, New York—San Francisco—London, 1978, Chapt. 2, p. 53–109
     b) Gokel, G. W., Korzeniowsky, S.: "Macrocyclic Polyether Syntheses", (Hafner, K., Rees, C. W., Trost, B. M., Lehn, J.-M., von Ragué Schleyer, P., Zahradnik, R., eds.), Springer-Verlag, Berlin—Heidelberg—New York, 1982
119. Review: Bradshaw, J. S., Maas, G. E., Izatt, R. M., Christensen, J. J.: Chem. Rev. *79*, 37 (1979)
120. Review: Bradshaw, J. S., Stott, P. E.: Tetrahedron *36*, 461 (1980)
121. Williamson, A. W.: J. Chem. Soc. *4*, 106, 229 (1852)
122. Weber, E., Vögtle, F.: Inorg. Chim. Acta *45*, L65 (1980)
123. Weber, E., Vögtle, F.: Chem. Ber. *109*, 1803 (1976); see also: Tietze, L.-F., Eicher, T.: Reaktionen und Synthesen im organisch-chemischen Praktium, Georg Thieme Verlag, Stuttgart—New York 1981, p. 328
124. Vögtle, F., Zuber, M.: Tetrahedron Lett. *1972*, 561
125. a) Weber, E.: Angew. Chem. *91*, 230 (1979); Angew. Chem., Int. Ed. Engl. *18*, 219 (1979)
     b) Weber, E.: J. Org. Chem. *47*, 3478 (1982)
     c) Cf. Weber, E.: Chem. Ber. *114*, 1551 (1981)
126. Review: Schill, G. in: "Catenanes, Rotaxanes and Knots", Academic Press, New York—London 1971
127. a) Müller, A., Sauerwald, A.: Monaths. Chem. *48*, 727 (1927)
     b) Müller, A., Bleier, P.: ibid. *56*, 391 (1930)
128. von Braun, J., Goll, O.: Ber. Dtsch. Chem. Ges. *60*, 1533 (1927)
129. Ziegler, K., Orth, P.: ibid. *66*, 1867 (1933); cf. also ref. [16]
130. Marckwald, W.: ibid. *31*, 3264 (1898)
131. Müller, A., Kindlmann, L.: ibid. *74*, 416 (1941)
132. a) Cf. Ruggli, P.: Liebigs Ann. Chem. *399*, 174 (1913)
     b) Cf. Ruggli, P.: ibid. *412*, 1 (1917)
     c) Cf. ref. [5]
133. See also references [106, 107]
134. Müller, A., Šrepel, E., Funder-Fritzsche, E., Dicher, F.: Monatsh. Chem. *83*, 386 (1952)
135. Stetter, H., Roos, E.-E.: Chem. Ber. *87*, 566 (1954)
136. Cf. Howard, C. C., Marckwald, W.: Ber. Dtsch. Chem. Ges. *32*, 2041 (1899)
137. Wiesner, K., Orr, D. E.: Tetrahedron Lett. *1960*, 11
138. a) Cf. Stetter, H., Mayer, K.-H.: Chem. Ber. *94*, 1410 (1961)
     b) Wehner, W., Vögtle, F.: Chem. Exp. Didakt. *1*, 77 (1975)
139. Doornbos, T., Strating, J.: Org. Prep. Proc. *1970*, 101
140. a) Fuson, R. C., House, H. O.: J. Am. Chem. Soc. *75*, 1327 (1953)
     b) Fuson, R. C., House, H. O.: ibid. *75*, 5744 (1953)
141. a) Hama, F., Sakata, Y., Misumi, S.: Tetrahedron Lett. *1981*, 1123
     b) Doyama, K., Hama, F., Sakata, Y., Misumi, S.: ibid. *1981*, 4101
142. Lüttringhaus, A., Simon, H.: Liebigs Ann. Chem. *557*, 120 (1945)

143. Stetter, H., Roos, E.-E.: Chem. Ber. *88*, 1390 (1955)
144. Stetter, H.: Chem. Ber. *86*, 197 (1953)
145. Stetter, H.: ibid. *86*, 380 (1953)
146. Odashima, K., Itai, A., Iitaka, Y., Koga, K.: J. Am. Chem. Soc. *102*, 2504 (1980)
147. Schill, G., Lüttringhaus, A.: Angew. Chem. *76*, 567 (1964); Angew. Chem., Int. Ed. Engl. *3*, 546 (1964)
148. a) Schill, G.: Chem. Ber. *98*, 2906 (1965)
    b) Schill, G.: ibid. *99*, 2689 (1966)
    c) Schill, G.: ibid. *100*, 2021 (1967)
    d) Schill, G.: Liebigs Ann. Chem. *695*, 65 (1966)
149. a) Schill, G., Murjahn, K., Beckmann, W.: Chem. Ber. *105*, 3591 (1972)
    b) Schill, G., Zürcher, C.: Naturwissenschaften *58*, 40 (1971)
150. Schill, G., Ortlieb, H.: Chem. Ber. *114*, 877 (1981)
151. a) Schill, G., Zürcher, C.: Angew. Chem. *81*, 996 (1969); Angew. Chem., Int. Ed. Engl. *8*, 988 (1969)
    b) Schill, G., Zürcher, C.: Chem. Ber. *110*, 2046 (1977)
    c) Schill, G., Zollenkopf, H.: Liebigs Ann. Chem. *721*, 53 (1969)
    d) Schill, G., Zürcher, C., Vetter, W.: Chem. Ber. *106*, 228 (1973)
    e) Schill, G., Rißler, K., Fritz, H., Vetter, W.: Angew. Chem. *93*, 197 (1981); Angew. Chem., Int. Ed. Engl. *20*, 187 (1981)
152. Vögtle, F., Neumann, P.: Tetrahedron Lett. *1970*, 115
153. Vögtle, F.: Tetrahedron Lett. *1968*, 3623
154. Vögtle, F., Neumann, P.: Chem. Commun. *1970*, 1464
155. Knipe, A. C., Stirling, C. J. M.: J. Chem. Soc. [B] *1968*, 67
156. Schill, G., Rothmaier, K., Ortlieb, H.: Synthesis *1972*, 426
157. a) Powers, J. C., Seidner, R., Parsons, T. G.: Tetrahedron Lett. *1965*, 1713
    b) Brimacombe, J. S., Jones, B. D., Stacey, M., Willerd, J. J.: Carbohydr. Res. *2*, 167 (1966)
    c) Nasipuri, D., Bhattacharyya, A., Hazra, B. G.: Chem. Commun. *1971*, 660
    d) Fieser, L. F., Fieser, M.: "Reagents for Organic Synthesis", Vol. I, J. Wiley, New York, 1967, p. 278
158. Tabushi, I., Kobuke, Y., Ando, K., Kishimoto, M., Ohara, E.: J. Am. Chem. Soc. *102*, 5947 (1980)
159. Casadei, M. A., Galli, C., Mandolini, L.: J. Org. Chem. *46*, 3127 (1981)
160. Griffin, Jr., R. W., Coburn, R. A.: Tetrahedron Lett. *1964*, 2571
161. Vögtle, F., Zuber, M.: Synthesis *1972*, 543
162. Shinmyozu, T., Inazu, T., Yoshino, T.: Chem. Lett. *1976*, 1405
163. a) Ruggli, P., Staub, A.: Helv. Chim. Acta *20*, 918 (1937)
    b) Kipping, J. S.: Ber. Dtsch. Chem. Ges. *28*, 30 (1888)
164. a) Shinmyozu, T., Inazu, T., Yoshino, T.: Chem. Lett. *1977*, 1347
    b) Shinmyozu, T., Inazu, T., Yoshino, T.: ibid. *1978*, 1319
    c) Shinmyozu, T., Inazu, T., Yoshino, T.: ibid. *1979*, 541
165. Review: Asinger, F., Vogel, H. H.: „Herstellung von Alkanen und Cycloalkanen", in: Houben-Weyl-Müller (Methoden der Organischen Chemie), 4th ed., vol. V/1a, p. 347, Thieme Verlag, Stuttgart 1970
166. a) Wurtz, A.: Liebigs Ann. Chem. *96*, 364 (1855)
    b) Wurtz, A.: Ann. Chim. Phys. [3] *44*, 275 (1855)
167. Baker, W., Banks, R., Lyon, D. R., Mann, F. G.: J. Chem. Soc. *1945*, 27
168. Cram, D. J., Steinberg, H.: J. Am. Chem. Soc. *73*, 5691 (1951)
169. Baker, W., McOmie, J. F. W., Norman, J. M.: J. Chem. Soc. *1951*, 1114
170. Lindsay, W. S., Stokes, P., Humber, L. G., Boekelheide, V.: J. Am. Chem. Soc. *83*, 943 (1961)
171. Cope, A. C., Fenton, S. W.: ibid. *73*, 1668 (1951)
172. Müller, E., Röscheisen, G.: Chem. Ber. *90*, 543 (1957)
173. Boekelheide, V., Phillips, J. B.: J. Am. Chem. Soc. *89*, 1695 (1967)
174. Paioni, R., Jenny, W.: Helv. Chim. Acta *53*, 141 (1970)
175. a) Paioni, R., Jenny, W.: ibid. *52*, 2041 (1969)
    b) Jenny, W., Holzrichter, H.: Chimia *22*, 306 (1968)
    c) Jenny, W., Holzrichter, H.: ibid. *22*, 247 (1968)

d) Jenny, W., Holzrichter, H.: ibid. *22*, 139 (1968)

e) Jenny, W., Paioni, R.: ibid. *22*, 142 (1968)

f) Jenny, W., Burri, K.: ibid. *21*, 186 (1967)

g) Jenny, W., Holzrichter, H.: ibid. *21*, 509 (1967)

h) Jenny, W., Holzrichter, H.: ibid. *23*, 158 (1969)

176. Burri, K., Jenny, W.: Helv. Chim. Acta *50*, 1978 (1967)

177. Bergmann, E. D., Pelchowicz, Z.: J. Am. Chem. Soc. *75*, 4281 (1953)

178. Flammang, R., Figeys, H. P., Martin, R. H.: Tetrahedron *24*, 1171 (1968)

179. Sato, T., Akabori, S., Muto, S., Hata, K.: ibid. *24*, 5557 (1968)

180. Gilman, H., Beel, J. A., Brannen, C. G., Bullock, M. W., Dunn, G. E., Miller, L. S.: J. Am. Chem. Soc. *71*, 1499 (1949)

181. Allinger, N. L., Da Rooge, M. A., Hermann, R. B.: ibid. *83*, 1974 (1961)

182. Baker, W., Buggle, K. M., McOmie, J. F. W., Watkins, D. A. M.: J. Chem. Soc. *1958*, 3594

183. Boekelheide, V., Pepperdine, W.: J. Am. Chem. Soc. *92*, 3684 (1970)

184. Peter, R., Jenny, W.: Helv. Chim. Acta *49*, 2123 (1966)

185. Vögtle, F., Staab, H. A.: Chem. Ber. *101*, 2709 (1968)

186. Bieber, W., Vögtle, F.: ibid. *112*, 1919 (1979)

187. Only a few thiapolyether esters have been synthesized through C—S-bond formation. They are mentioned in section III.1.1.2; see also references [204, 207, 208, 210, 211]

188. Woodward, R. B.: Angew. Chem. *69*, 50 (1957)

189. Review: Masamune, S., Bates, G. S., Corcoran, J. W.: Angew. Chem. *89*, 602 (1977); Angew. Chem., Int. Ed. Engl. *16*, 585 (1977)

190. Crawford, L. M. R., Drewes, S. E., Sutton, D. A.: Chem. Ind. (London) *1970*, 1351

191. Vögtle, F., Lichtenthaler, R.: Chem.-Ztg. *94*, 727 (1970)

192. Corey, E. J., Nicolaou, K. C.: J. Am. Chem. Soc. *96*, 5614 (1974)

193. Review: Back, T. G.: Tetrahedron *33*, 3041 (1977)

194. Masamune, S.: Aldrichimica Acta *11*, 23 (1978)

195. Some important macrolides are listed in reference [189], pp. 618, 619

196. a) Stoll, M., Rouvé, A.: Helv. Chim. Acta *17*, 1283 (1934)

    b) Stoll, M., Rouvé, A.: ibid. *18*, 1087 (1935)

197. a) Stoll, M., Bolle, P.: ibid. *31*, 98 (1948)

    b) Collaud, C.: ibid. *25*, 965 (1942)

198. Taub, D., Girotra, N. N., Hoffsommer, R. D., Kuo, C. H., Slates, H. L., Weber, S., Wendler, N. L.: Tetrahedron *24*, 2443 (1968)

199. Staab, H. A.: Angew. Chem. *74*, 407 (1962); Angew. Chem., Int. Ed. Engl. *1*, 351 (1962)

200. Bates, G. S., Diakur, J., Masamune, S.: Tetrahedron Lett. *1976*, 4423

201. Kerschbaum, M.: Ber. Dtsch. Chem. Ges. *60*, 902 (1927)

202. Izatt, R. M., Lamb, J. D., Maas, G. E., Asay, R. E., Bradshaw, J. S., Christensen, J. J.: J. Am. Chem. Soc. *99*, 2365 (1977)

203. Izatt, R. M., Lamb, J. D., Asay, R. E., Maas, G. E., Bradshaw, J. S., Christensen, J. J., Moore, S. S.: ibid. *99*, 6134 (1977)

204. Frensch, K., Vögtle, F.: Tetrahedron Lett. *1977*, 2573

205. Frensch, K., Vögtle, F.: J. Org. Chem. *44*, 884 (1979)

206. Bradshaw, J. S., Thompson, M. D.: ibid. *43*, 2456 (1978)

207. a) For the synthesis of thio esters also THF is used

    b) Matsushima, K., Kawamura, N., Okahara, M.: Tetrahedron Lett. *1979*, 3445

    c) Nakatsuh, Y., Kawamura, N., Okahara, M.: Synthesis *1981*, 42

208. Review: Fore, P. E., Bradshaw, J. S., Nielsen, S. F.: J. Heterocycl. Chem. *15*, 269 (1978)

209. Review: Thompson, M. D., Bradshaw, J. S., Nielsen, S. F., Bishop, C. T., Cox, F. T., Fore, P. E., Maas, G. E., Izatt, R. M., Christensen, J. J.: Tetrahedron *33*, 3317 (1977)

210. Frensch, K., Oepen, G., Vögtle, F.: Liebigs Ann. Chem. *1979*, 858

211. Bradshaw, J. S., Bishop, C. T., Nielsen, S. F., Asay, R. E., Mashidas, D. R. K., Flanders, E. D., Hansen, L. D., Izatt, R. M., Christensen, J. J.: J. C. S. Perkin I, *1976*, 2505

212. Review: Bradshaw, J. S., Asay, R. E., Maas, G. E., Izatt, R. M., Christensen, J. J.: J. Heterocycl. Chem. *15*, 825 (1978)

213. Taub, D., Girotra, N. N., Hoffsommer, R. D., Kuo, C. H., Slates, H. L., Weber, S., Wendler, N. L.: Chem. Commun. *1967*, 225

214. Wehrmeister, H. L., Robertson, D. E.: J. Org. Chem. *33*, 4173 (1968)

215. White, J. D., Lodwig, S. N., Trammell, G. L., Fleming, M. P.: Tetrahedron Lett. *1974*, 3263
216. Wrobel, J. T., Golebiewski, W. M.: ibid. *1973*, 4293
217. Girotra, N. N., Wendler, N. L.: Chem. Ind. (London) *1967*, 1493
218. Sakamoto, K., Oki, M.: Chem. Lett. *1975*, 645
219. Sakamoto, K., Oki, M.: Tetrahedron Lett. *1973*, 3989
220. Zahn, H., Repin, J. F.: Chem. Ber. *103*, 3041 (1970)
221. Meraskentis, E., Zahn, H.: ibid. *103*, 3034 (1970)
222. Stetter, H., Marx, J.: Liebigs Ann. Chem. *607*, 59 (1957)
223. Rothe, M.: Angew. Chem. *74*, 725 (1962); Angew. Chem., Int. Ed. Engl. *1*, 669 (1962)
224. Rothe, M., Rothe, I., Brünig, H., Schwenke, K. D.: Angew. Chem. *71*, 700 (1959)
225. Nagao, Y., Seno, K., Miyasaka, T., Fujita, E.: Chem. Lett. *1980*, 159
226. See ref. [138a)]
227. Simmons, H. E., Park, C. H.: J. Am. Chem. Soc. *90*, 2428 (1968)
228. Buhleier, E., Wehner, W., Vögtle, F.: Synthesis *1978*, 155
229. Schneider, F., Bernauer, K., Guggisberg, A., van den Broek, P., Hesse, M., Schmid, H.: Helv. Chim. Acta *57*, 434 (1974)
230. Guggisberg, A., van den Broek, P., Hesse, M., Schmid, H., Schneider, F., Bernauer, K.: ibid. *59*, 3013 (1976)
231. Schmidtchen, F. P.: Chem. Ber. *113*, 864 (1980)
232. Vögtle, F., Dix, J. P.: Liebigs Ann. Chem. *1977*, 1698
233. Dietrich, B., Lehn, J.-M., Sauvage, J. P., Blanzat, J.: Tetrahedron *29*, 1629 (1973)
234. a) Buhleier, E., Wehner, W., Vögtle, F.: Chem. Ber. *111*, 200 (1978)
     b) Buhleier, E., Wehner, W., Vögtle, F.: ibid. *112*, 546 (1979)
235. Buhleier, E., Wehner, W., Vögtle, F.: Liebigs Ann. Chem. *1978*, 537
236. a) Wester, N., Vögtle, F.: Chem. Ber. *112*, 3723 (1979)
     b) Wester, N., Vögtle, F.: ibid. *113*, 1487 (1980)
     c) Wester, N., Vögtle, F.: J. Chem. Res. (S) *1978*, 400; (M) *1978*, 4856
     d) Sieger, H., Vögtle, F.: Liebigs Ann. Chem. *1980*, 425
     e) Oepen, G., Vögtle, F.: ibid. *1979*, 1094
237. Lehn, J.-M., Simon, J., Wagner, J.: Angew. Chem. *85*, 621 (1973); Angew. Chem., Int. Ed. Engl. *12*, 578 (1973)
238. Dietrich, B., Lehn, J.-M., Sauvage, J. P.: Chem. Commun. *1970*, 1055
239. Ando, N., Yamamoto, Y., Oda, J., Inouye, Y.: Synthesis *1978*, 688
240. Tabushi, I., Okino, H., Kuroda, Y.: Tetrahedron Lett. *1976*, 4339
241. See also references cited in the reviews [11, 29, 109–111, 116, 119, 120)]
242. Stetter, H., Marx-Moll, L.: Chem. Ber. *91*, 677 (1958)
243. a) Sakamoto, K., Oki, M.: Bull. Chem. Soc. Jpn. *46*, 270 (1973)
     b) Kondo, H., Okamoto, H., Jun-ichi Kikuchi, Sunamoto, J.: J. C. S. Perkin I, *1981*, 3125
     c) Buhleier, E., Wehner, W., Vögtle, F.: Chem. Ber. *112*, 559 (1979)
244. Review: Stewart, W. E., Siddall, T.: Chem. Rev. *70*, 517 (1970)
245. Urushigawa, Y., Inazu, T., Yoshino, T.: Bull. Chem. Soc. Jpn. *44*, 2546 (1971)
246. Doornbos, T., Strating, J.: Synth. Commun. *1971*, 11
247. Dittmer, C. D., Blidner, B. B.: J. Org. Chem. *38*, 2873 (1973)
248. Murakami, Y., Nakano, A., Miyata, R., Matsuda, Y.: J. C. S. Perkin I, *1979*, 1669
249. See ref. [235)]
250. Overmann, L. E.: J. Org. Chem. *37*, 4214 (1972)
251. Rossa, L., Vögtle, F.: Liebigs Ann. Chem. *1981*, 459
252. a) Schmidt, U., Griesser, H., Lieberknecht, A., Talbiersky, J.: Angew. Chem. *93*, 271 (1981); Angew. Chem., Int. Ed. Engl. *20*, 280 (1981)
     b) Schmidt, U., Lieberknecht, A., Griesser, H., Häusler, J.: Angew. Chem. *93*, 272 (1981); Angew. Chem., Int. Ed. Engl. *20*, 281 (1981)
253. March, J.: "Advanced Organic Chemistry", Reactions, Mechanisms and Structure, 2nd ed., McGraw-Hill, Kogakusha, 1977, pp. 444
254. Review: Schaefer, J. P., Bloomfield, J. J.: Org. React. (Adams, R., Blatt, A. H., Boekelheide, V., Clairns, T. L., Cope, A. C., Cram, D. J., House, H. O., eds.), Vol. 15, Wiley 1967
255. Leonard, N. J., Schimelpfenig, Jr., C. W.: J. Org. Chem. *23*, 1708 (1958)
256. Blicke, F. F., Azuara, J., Doorenbos, N. J., Hotelling, E. B.: J. Am. Chem. Soc. *75*, 5418 (1953)
257. Leonard, N. J., Sentz, R. C.: J. Am. Chem. Soc. *74*, 1704 (1952)

258. Leonard, N. J., Adamcik, J. A., Djerassi, C., Halpern, O.: ibid. *80*, 4858 (1958)
259. Leonard, N. J., Oki, M.: ibid. *77*, 6241 (1955)
260. Leonard, N. J., Milligan, T. W., Brown, T. L.: ibid. *82*, 4075 (1960)
261. Lüttringhaus, A., Prinzbach, H.: Liebigs Ann. Chem. *624*, 79 (1959)
262. Leonard, N. J., Oki, M., Chiavarelli, S.: J. Am. Chem. Soc. *77*, 6234 (1955)
263. Schimelpfenig, C. W., Lin, Y.-T., Walter, Jr., J. F.: J. Org. Chem. *28*, 805 (1963)
264. Mock, W., Richards, J. H.: ibid. *27*, 4050 (1962)
265. Hurd, R. N., Shah, D. H.: ibid. *38*, 390 (1973)
266. Review: Schäfer, J. P., Bloomfield, J. J.: Org. React. *15*, 28 (1967)
267. See ref. [253)], p. 873, pp. 1137–1138
268. a) Moore, C. W., Thorpe, J. F.: J. Chem. Soc. *93*, 165 (1908)
     b) Baron, H., Remfry, F. G. P., Thorpe, J. F.: J. Chem. Soc. *85*, 1726 (1904)
269. Ziegler, K., Lüttringhaus, A.: Liebigs Ann. Chem. *511*, 1 (1934)
270. Ziegler, K., Eberle, H., Ohlinger, H.: ibid. *504*, 94 (1933)
271. Ziegler, K., Aurnhammer, R.: ibid *513*, 43 (1934)
272. Ziegler, K., Weber, K.: ibid. *512*, 164 (1934)
273. Allinger, N. L., Greenberg, S.: J. Am. Chem. Soc. *84*, 2394 (1962)
274. Nitzschke, H.-J., Budka, H.: Chem. Ber. *88*, 264 (1955)
275. Cope, A. C., Mehta, A. S.: J. Am. Chem. Soc. *86*, 1268 (1964)
276. Cope, A. C., Cotter, R. J.: J. Org. Chem. *29*, 3467 (1964)
277. Nitzschke, H.-J., Faerber, G.: Chem. Ber. *87*, 1635 (1954)
278. Muth, C. W., Steiniger, D. O., Papanastassiou, Z. B.: J. Am. Chem. Soc. *77*, 1006 (1955)
279. Fry, E. M., Fieser, L. F.: ibid. *62*, 3489 (1940)
280. Baldwin, S.: J. Org. Chem. *26*, 3280, 3288 (1961)
281. See ref. [253)], pp. 864
282. Review: Vollhardt, K. P. C.: Synthesis *1975*, 765
283. Wittig, G., Eggers, H., Duffner, P.: Liebigs Ann. Chem. *619*, 10 (1958)
284. Grohmann, K., Howes, P. D., Mitchell, R. H., Monahan, A., Sondheimer, F.: J. Org. Chem. *38*, 808 (1973)
285. Staab, H. A., Wehinger, E., Thorwart, W.: Chem. Ber. *105*, 2290 (1972)
286. Bestmann, H. J., Zimmermann, R.: Fortschr. Chem. Forsch. *20*, 1 (1971)
287. Schlosser, M.: in „Topics in Stereochemistry" (Eliel, E. L., Allinger, N. L., eds.), Vol. 5, Interscience, New York 1970, p. 1
288. a) Reucroft, J., Sammes, P. G.: Q. Rev. Chem. Soc. *25*, 137 (1977)
     b) Fuji, K., Ichikawa, K., Fujita, E.: J. C. S. Perkin I, *1980*, 1066
289. Boutagy, J., Thomas, R.: Chem. Rev. *74*, 87 (1974)
290. Review: Nicolaou, K. C.: Tetrahedron *33*, 683 (1977)
291. Wadsworth, Jr., W. S., Emmons, W. D.: J. Am. Chem. Soc. *83*, 1733 (1961)
292. Arbuzov, B. A.: Pure Appl. Chem. *9*, 307 (1964)
293. Stork, G., Nakamura, E.: J. Org. Chem. *44*, 4010 (1979)
294. Nicolaou, K. C., Seitz, S. P., Pavia, M. R., Petasis, N. A.: ibid. *44*, 4011 (1979)
295. a) Burri, K. F., Cardone, R. A., Wen Yean Chen, Rosen, P.: J. Am. Chem. Soc. *100*, 7069 (1978)
     b) Floyd, D. M., Fritz, A. W.: Tetrahedron Lett. *1981*, 2847
296. Review: Bloomfield, J. J., Owsley, D. C., Nelke, J. M. in: Org. React. Vol. *23* (Baldwin, J. E., Heck, R. F., Kende, A. S., Leimgruber, W., Marshall, J. A., McKusik, B. C., Meinwald, J., Trost, B. M., eds.), pp. 259 (1976), John Wiley & Sons, Inc., New York—London—Sydney—Toronto. Cf. references cited therein
297. Finley, K. T.: Chem. Rev. *64*, 573 (1964)
298. Corey, E. J.: Pure Appl. Chem. *14*, 19 (1967)
299. Allinger, N. L.: Org. Synth. Coll. Vol. IV, Wiley, New York, N.Y. 1963, p. 840
300. Johnson, P. Y., Zitsman, J., Hatch, C. E.: J. Org. Chem. *38*, 4087 (1973)
301. Park, C. H., Simmons, H. E.: J. Am. Chem. Soc. *94*, 7184 (1972)
302. Finley, K. T., Sasaki, N. A.: ibid. *88*, 4267 (1966)
303. Cram, D. J., Gaston, L. K.: ibid. *82*, 6386 (1960)
304. Blomquist, A. T., Meinwald, Y. C.: ibid. *80*, 630 (1958)
305. Cram, D. J., Allinger, N. L.: ibid. *78*, 2518 (1956)
306. Cope, A. C., Fenton, S. W., Spencer, C. F.: ibid. *74*, 5884 (1952)

307. Knight, J. D., Cram, D. J.: ibid. *73*, 4136 (1951)
308. Murakami, Y., Aoyama, Y., Kida, M., Nakano, A., Dobashi, K., Chieu Dinh Tran, Matsuda, Y.: J. C. S. Perkin I, *1979*, 1560
309. Mislow, K., Hyden, S., Schaefer, H.: Tetrahedron Lett. *1961*, 410
310. Cram, D. J., Allinger, N. L., Steinberg, H.: J. Am. Chem. Soc. *76*, 6132 (1954)
311. See ref. [253], pp. 1134
312. Bloomfield, J. J.: Tetrahedron Lett. *1968*, 591
313. Goldfarb, Ya. R., Taits, S. Z., Belen'kii, L. I.: Tetrahedron *19*, 1851 (1963)
314. Kober, A. E., Westmann, T. L.: J. Org. Chem. *35*, 4161 (1970)
315. Rühlmann, K.: Synthesis *1971*, 236 and references cited therein
316. Schräpler, U., Rühlmann, K.: Chem. Ber. *97*, 1383 (1964)
317. Strating, J., Reiffers, S., Wynberg, H.: Synthesis *1971*, 209
318. a) Glaser, C.: Ber. Dtsch. Chem. Ges. *2*, 422 (1869)
     b) Glaser, C.: Liebigs Ann. Chem. *137*, 154 (1870)
319. a) Chodkiewicz, W., Cadiot, P.: C. R. Acad. Sci. *241*, 1055 (1955)
     b) Chodkiewicz, W.: Ann. Chim. (Paris) [13] *2*, 819 (1957)
320. Eglinton, G., Galbraith, A. R.: Chem. Ind. (London) *1956*, 737
321. Eglinton, G., McCrae, W.: Adv. Org. Chem. *4*, 225 (1963); cf. references cited therein
322. Carnduff, J., Eglinton, G., McCrae, W., Raphael, R. A.: Chem. Ind. (London) *1960*, 559
323. Meier, H.: Synthesis *1972*, 235; cf. references cited therein
324. Eglinton, G., Galbraith, A. R.: J. Chem. Soc. *1959*, 889
325. Bergelson, L. D., Molotkovsky, J. G., Shemyakin, M. M.: Chem. Ind. (London) *1960*, 558
326. Ojima, J., Shiroishi, Y., Wada, K., Sondheimer, F.: J. Org. Chem. *45*, 3564 (1980)
327. Behr, O. M., Eglinton, G., Galbraith, A. R., Raphael, R. A.: J. Chem. Soc. *1960*, 3614
328. Behr, O. M., Eglinton, G., Raphael, R. A.: Chem. Ind. (London) *1959*, 699
329. Behr, O. M., Eglinton, G., Lardy, I. A., Raphael, R. A.: J. Chem. Soc. *1964*, 1151
330. Matsuoka, T., Sakata, Y., Misumi, S.: Tetrahedron Lett. *1970*, 2549
331. Howes, P. D., LeGoff, E., Sondheimer, F.: ibid. *1972*, 3691
332. Staab, H. A., Mack, H., Wehinger, E.: ibid. *1968*, 1465
333. Amiel, Y., Sondheimer, F., Wolovsky, R.: Proc. Chem. Soc. *1957*, 22
334. Sondheimer, F., Wolovsky, R.: J. Am. Chem. Soc. *84*, 260 (1962)
335. Sondheimer, F., Wolovsky, R., Amiel, Y.: ibid. *84*, 274 (1962)
336. Takahashi, T., Nagashima, T., Ikeda, H., Tsuji, J.: Tetrahedron Lett. *1982*, 4361

Syntheses and Properties of the [2$_n$]Cyclophanes

Virgil Boekelheide

Department of Chemistry, University of Oregon Eugene, Oregon, 97403 USA

Table of Contents

1 Introduction and Nomenclature

Although the first report on [2,2]paracyclophane (4) was its isolation from a polymer and its identity was established by an x-ray analysis [1–3], the development of cyclophane chemistry as a field of research is due to the pioneering work of Professor Cram and his students [4–6]. In the opening remarks of his first paper [7], Cram foresaw that points of interest in these molecules would be

1) "interstitial resonance effects"
2) effects on rates and unusual substituent directive effects as compared to benzene; and
3) the possibility of intramolecular (charge transfer) molecular complexes.

Rarely have such early chemical predictions been so completely realized by subsequent investigations.

The present review covers the syntheses and properties of [2ₙ]cyclophanes. Since Cram first applied the trivial name paracyclophane to describe (4) [7], phane nomenclature has evolved through use and necessity to become a reasonably efficient way of correlating a large body of rather complicated structures [8–13]. Under the rules of "phane" nomenclature [11], the term cyclophane is reserved for bridged benzene rings and the present name [2ₙ]cyclophanes relates to phanes having two benzene decks with two to six bridges.

In *Chart 1* are presented the twelve possible [2ₙ]cyclophanes in which the sub-

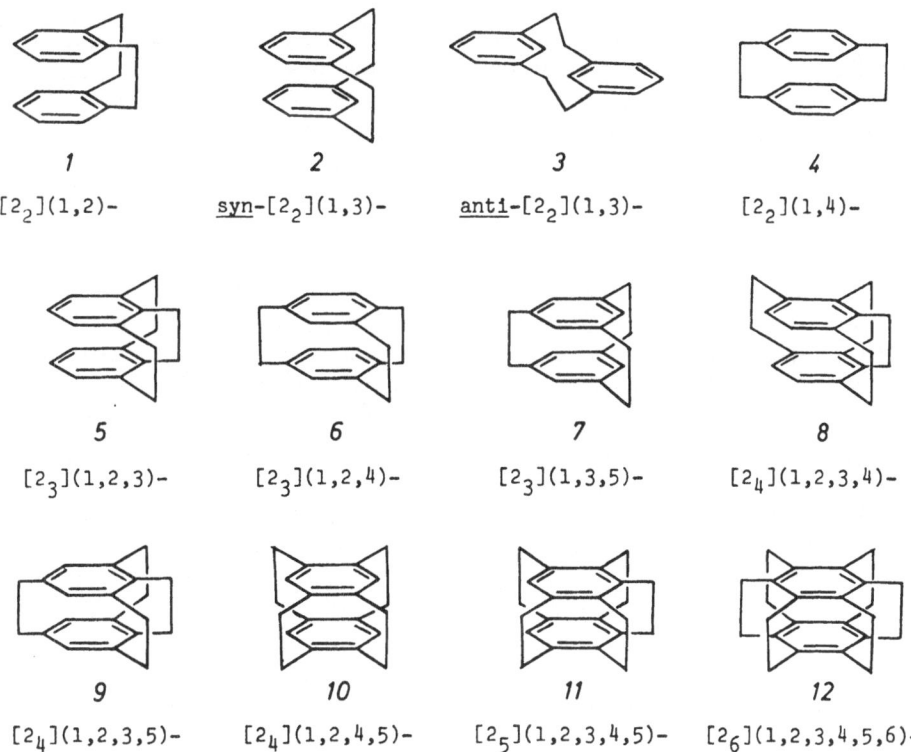

1	*2*	*3*	*4*
[2₂](1,2)–	syn-[2₂](1,3)–	anti-[2₂](1,3)–	[2₂](1,4)–
5	*6*	*7*	*8*
[2₃](1,2,3)–	[2₃](1,2,4)–	[2₃](1,3,5)–	[2₄](1,2,3,4)–
9	*10*	*11*	*12*
[2₄](1,2,3,5)–	[2₄](1,2,4,5)–	[2₅](1,2,3,4,5)–	[2₆](1,2,3,4,5,6)–

stitution pattern of the bridges in each deck is the same. In *Chart 2* are listed all of the known skewed [2ₙ]cyclophanes in which the substitution patterns of the bridges in the decks are not the same. How many more skewed [2ₙ]cyclophanes are possible is very difficult to predict. Our discussion will be limited to molecules of the carbon skeletons shown in *Charts 1* and *2*.

13	*14*	*15*
[2₂](1,3)(1,4)-	[2₃](1,2,4)(1,2,5)-	[2₃](1,2,4)(1,3,5)-

Currently, there is much concern with comparing the properties of the different [2ₙ]cyclophanes as these are affected by differing numbers of bridges and different bridge patterns. For these comparisons present nomenclature leads to certain ambiguities. It would be desirable to have a set procedure for naming and numbering each [2ₙ]cyclophane which would remain the same regardless of the presence or absence of substituents. As a part of this review, therefore, some refinements of the present system of nomenclature are proposed. These changes are proposed specifically for [2ₙ]cyclophanes but, quite possibility, it would be advantageous to extend them to phanes in general.

There are two points in the present system leading to ambiguity. One is the rule of following the longest path between bridgeheads with preference, in cases of equal choice, of giving a substituent the lower number. This requires that the numbering of the overall skeleton change depending on the presence or absence of substituents. It would be much more satisfactory to have a set system for numbering the skeletal framework and so avoid this ambiguity. Secondly, there is at present no set procedure for indicating bridging positions in skewed cyclophanes or in numbering bridges of [2ₙ]cyclophanes when more than two bridges are present. Thus, structure (16) could be named [2₄](1,2,4)-, [2₃](1,2,4)(1,3,4)-, or [2₃](1,2,4)(1,2,5)cyclophane (plus other variations depending on which parenthesis is first). To remove these difficulties the following refinements of the present nomenclature are proposed.

First, to make a decision about the numbers referring to bridgehead positions to be placed in parentheses, one looks down on the cyclophane molecule and in the top deck picks an anchor point bridgehead such that numbering around the ring in a *clockwise* manner from that anchor point will lead to the lowest possible number for the next bridgehead position. The resulting sequence of numbers for the bridgehead positions of the top deck is placed in the first parenthesis. Then, one goes to the lower deck and picks as the anchor point the bridgehead position which is directly linked to the anchor point of the top deck. Numbering around the bottom deck, again in a *clockwise* manner, gives a sequence of numbers for the bridgehead positions in the bottom deck and these are placed in the second parenthesis. When the bridgehead patterns are the same for both decks a second parenthesis is unnecessary.

To number the overal skeleton of the cyclophane it is proposed that one start, as in the present system, at the bridging atom attached to the anchor point of the lower deck and number up to the anchor point of the top deck. Numbering would continue around the ring of the top deck in a *clockwise* fashion. Then, from that stopping point one would proceed in a *clockwise* manner to the next bridgehead position and number the bridging atoms down to the lower deck. Next, the ring positions of the lower deck would be numbered following around in a *clockwise* fashion. Proceeding from that stopping point in the lower deck one would go in a *clockwise* fashion to the next bridgehead position where one would number the bridging atoms going to the top deck. Again, one would proceed in a clockwise fashion around the top deck to the next bridgehead position and number the bridging atoms down to the lower deck. This procedure for numbering bridging atoms would be continued until all of the skeletal atoms had been numbered.

For example, these refinements would lead in an unambiguous fashion to the assignment of names and numbering for structures (4), (16), and (12) as shown.

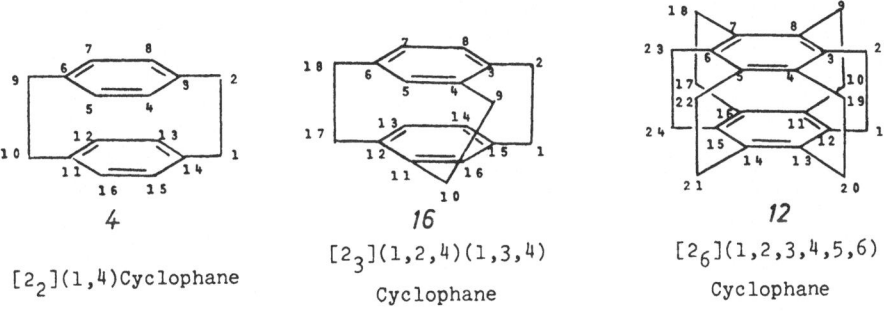

4

[2₂](1,4)Cyclophane

16

[2₃](1,2,4)(1,3,4)

Cyclophane

12

[2₆](1,2,3,4,5,6)

Cyclophane

2 Syntheses

2.1 Wurtz Coupling and 1,6-Elimination Reactions

In view of the number of good reviews [4-6,10,11,14-17] on cyclophanes, this report will tend to give brief treatment of earlier work and concentrate on more recent results. The Wurtz coupling reaction is the oldest of synthetic methods for cyclophanes, having been used by Pellegrin for the synthesis of *anti*-[2₂](1,3)cyclophane (3) in 1899 [18a]. Also, it was employed by Baker et al. for the synthesis of [2₂](1,2)-cyclophane (1) [18b], and by Cram and Steinberg for the synthesis of [2₂](1,4)cyclophane (4) [7]. Although the yields observed in the Wurtz coupling reaction are usually only about 20%, it is a useful method when the required dihalide is readily available.

The Hofmann-type, 1,6-elimination reaction of *p*-methylbenzylammonium hydroxides [19, 20] has been applied to the synthesis of a wide variety of [2₂](1,4)cyclophanes and is particularly valuable for preparing multiplayered cyclophanes [15]. Although yields in the 1,6-elimination reactions are usually low, Otsubo et al. have shown that they can be improved by careful attention to solvent, concentration, and inhibitor, as is shown for the preparation of (17) [21].

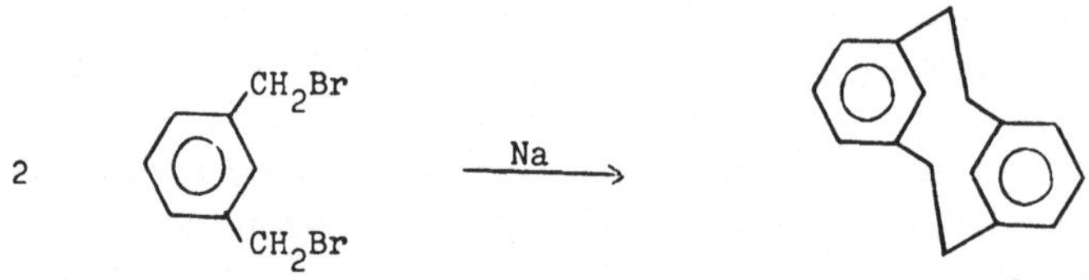

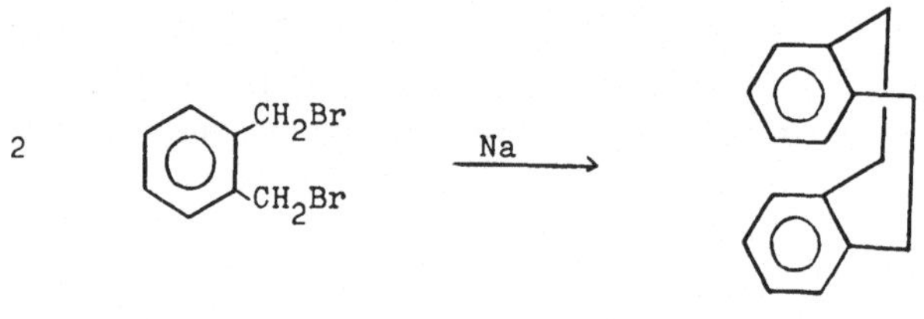

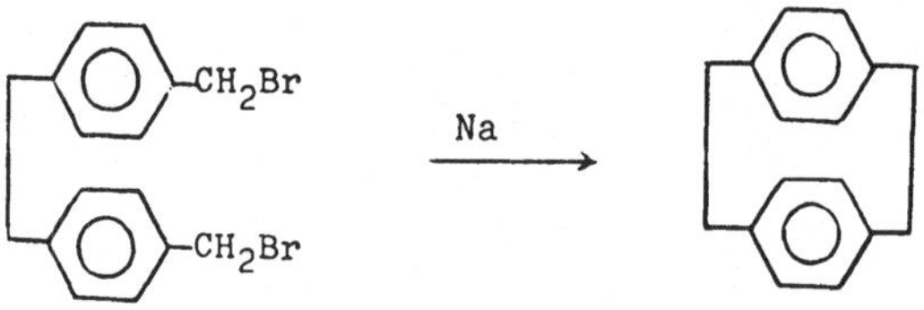

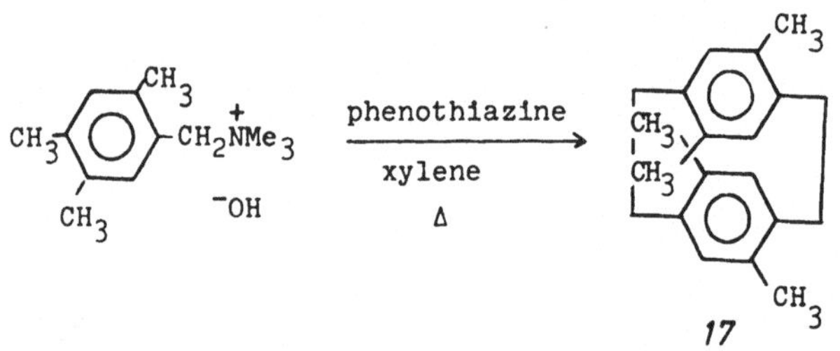

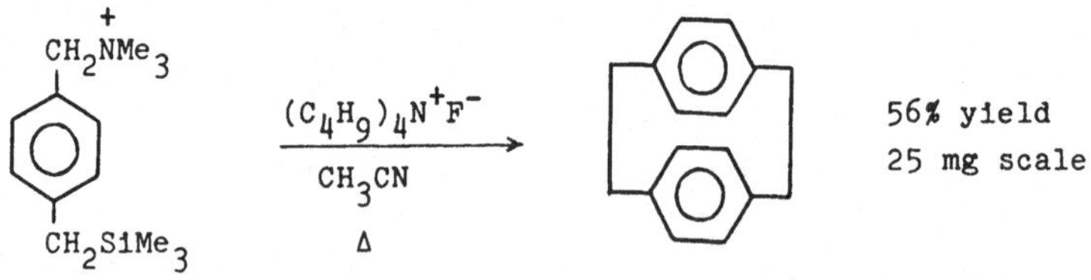

Recently, Ito et al. have synthesized (4) by an interesting variation involving fluoride ion attack on the trimethylsilyl analog (18) [22].

2.2 The Dithiacyclophane Approach

2.2.1 Coupling Reactions

During the period from 1951 to around 1970, a large number of cyclophanes were made using the Wurtz coupling reaction and the 1,6-elimination procedure. Then, introduction of methods for preparing dithiacyclophanes [23-33] and converting them via sulfur extrusion to cyclophanes [24, 26, 29, 30-33] led to a new and very general approach for making all types of cyclophanes. An example is the synthesis of 2,11-dithia-[3$_2$](1,4)cyclophane [28] and its conversion to [2$_2$](1,4)-cyclophane (4) in 75–80% yields [34].

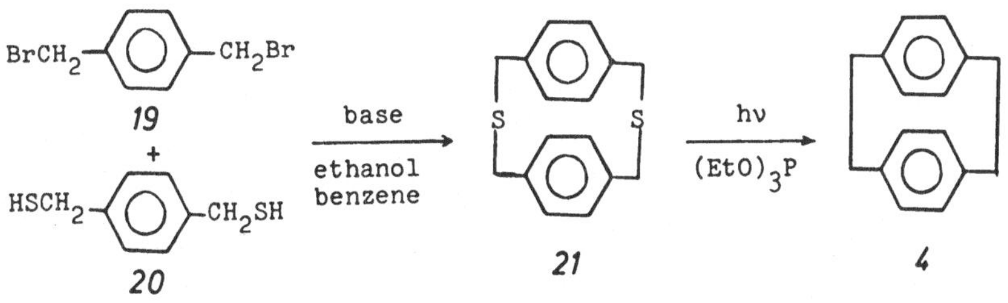

Although the coupling of dihalides with sodium sulfide to give dithiacyclophanes is successful [23], the yields of dithiacyclophanes are much better, usually about 80%, when a dihalide is coupled with a dimercaptan [25] in the presence of base under high dilution conditions [35]. In the case of the coupling of m-xylyl dihalides with m-xylyl dimercaptans, the presence of internal substituents [25] leads to the formation of both the syn and anti isomers, and these are not conformationally interconvertible. When both internal substituents are methyl, the ratio of the syn-(24) to anti-(25) isomers is 1:7 [33]. When one internal substituent is hydrogen and the other is methyl, only the anti-isomer (27) is isolated and it is not conformationally in equilibrium with (26) [36].

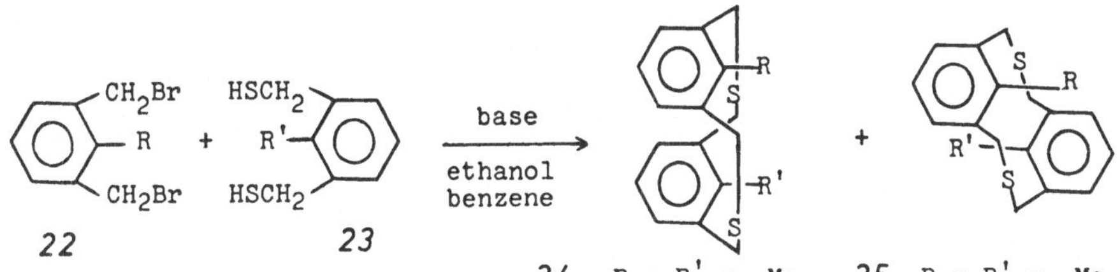

93

The relative amounts of the *syn-* and *anti*-isomers of the dithiacyclophanes formed in these coupling reactions is very dependent on the nature of the substituents. When one ring has electron donating groups, as in (28), and the other ring has electron-withdrawing groups, as in (29), the ratio of *syn-* to *anti*-isomers (30):(31) is 10:1 [37]. A possible explanation for this reversal of the relative proportions of *syn-* to *anti*-isomers is that a charge-transfer interaction bringing the benzene rings face to face in the transition state is quite important.

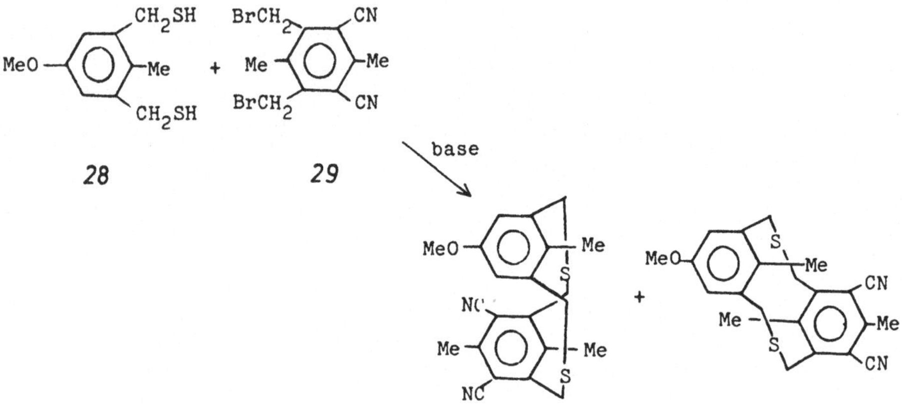

In the case where the internal substituents are hydrogen, the 2,11-dithia[3_2](1,3)-cyclophanes readily undergo conformational flipping between the *syn-* and *anti*-isomers. In the early work, it was assumed that the *anti*-isomer would be the more stable and would predominate in solution and be present in the crystalline state. This has proved not to be true. By a single crystal x-ray analysis, Anker, Bushnell and Mitchell showed that 2,11-dithia[3_2](1,3)cyclophane has the *syn*-conformation (32) in the crystalline state [38]. By an NMR spectral comparison, they concluded that 2,11-dithia[3_2](1,3)cyclophane also has the *syn*-conformation predominantly, if not exclusively, in solution. Furthermore, an extension of the NMR spectral comparison to a broad range of conformationally-mobile dithiacyclophanes shows that it is a common experience for the *syn*-conformation to be predominant.

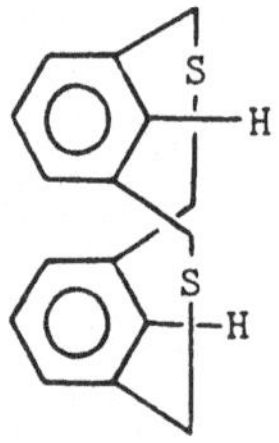

94

2.2.2 Elimination of Sulfur

Over the years a large number of methods have evolved for the removal of sulfur from dithiacyclophanes. These are summarized in Chart 3. The choice of method has differed depending on whether the ultimate goal was to prepare a cyclophane or a cyclophane-diene. In the beginning our interest was in the cyclophane-dienes (35) which, by valence tautomerization, give the 15,16-dialkyldihydropyrenes (36) [33].

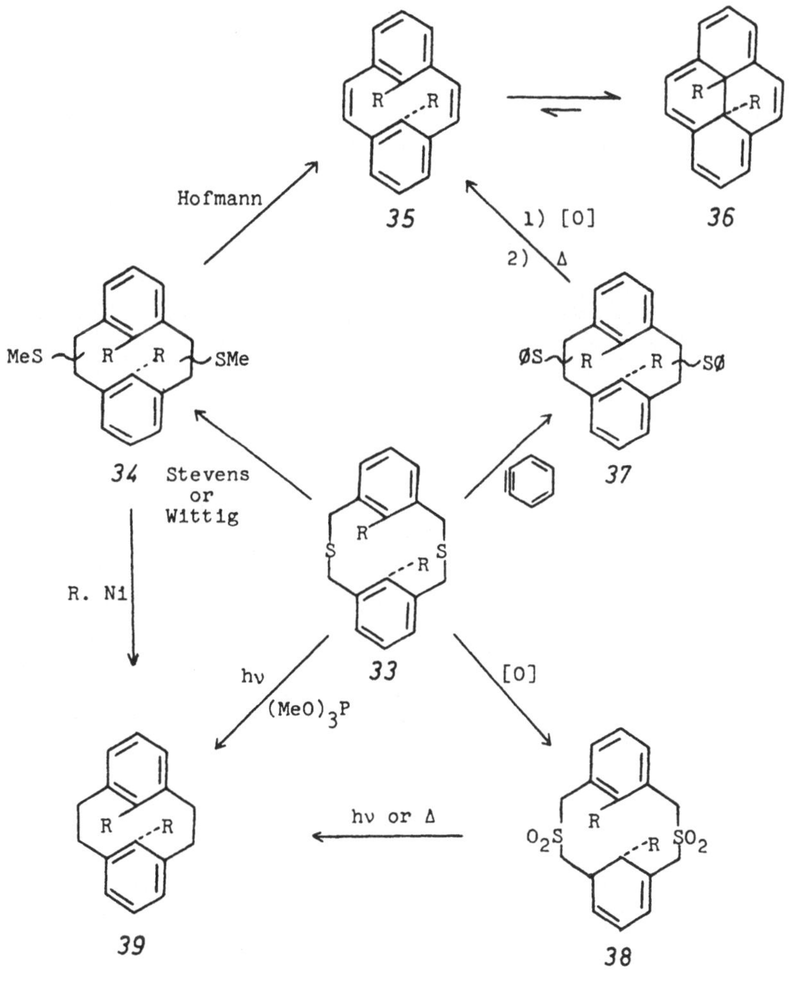

Ring contraction of the 2,11-dithia[3$_2$](1,3)cyclophanes (33) via the Stevens rearrangement [33], or the Wittig rearrangement [39], gives the corresponding disulfides (34), as a mixture of isomers, in high yield (>90%). Conversion of the disulfides (34) to the corresponding cyclophane-dienes (35) proceeds very well (>90%) by a Hofmann elimination in the metacyclophane series but often less well in the case of other cyclophanes. As alternatives to this step, therefore, the oxidation of the mixture of sulfides (34) to sulfoxides followed by pyrolysis has been developed [40]. Also, the conversion of the mixture of methyl sulfides (34) to the corresponding mixture of sulfones followed by treatment with base to effect elimination of methylsulfinic acid has been used successfully for preparing cyclophanes with unsaturated bridges [41].

Another alternative has been to carry out a modified Stevens rearrangement by treating the dithiacyclophane (33) with benzyne generated in situ to give the corresponding mixture of phenyl sulfides (37). Oxidation of the phenyl sulfides (37) to the corresponding sulfoxides followed by pyrolytic elimination of phenylsulfinic acid in boiling xylene then provides the cyclophane-dienes [42, 43]. This is the method of choice for preparing [2$_2$](1,4)cyclophane-1,9-diene (42) [42].

40	41	42
	65%	54%

For the conversion of dithiacyclophanes to cyclophanes with saturated bridges, there are again a variety of methods. The sulfides from the Stevens or Wittig rearrangements ((34) or (37)) can be desulfurized by treatment with Raney nickel to give the cyclophane (39) [33, 43, 44]. Irradiation of dithiacyclophanes (33) in the presence of trimethyl phosphite gives the corresponding cyclophanes (39), usually in excellent yields [34, 45]. Likewise, the conversion of dithiacyclophanes (33) to the corresponding bissulfones (38) followed by pyrolysis in the gas phase at 500 to 600 °C gives the corresponding cyclophanes (39), again usually in excellent yields [24, 26, 29]. An apparatus for carrying out these gas phase pyrolyses has been described [46]. Finally, the bissulfones (38) can be converted to the corresponding cyclophanes by irradiation to effect loss of sulfur dioxide [47-49].

It should be noted that pyrolyses of diesters, in analogy to pyrolyses of bissulfones, has been reported to give [2$_2$](1,4)cyclophane derivatives in excellent yield [50].

By use of the dithiacyclophane-sulfur extrusion route, examples of the cyclophanes shown by structures (4) [34], (3) [45], (2) [37], (7) [51] and (10) [52, 53], shown in Chart 1, and structures (13) [54], (14) [55], and (15) [55], shown in Chart 2, were prepared. For the tetra-bridged [2$_4$](1,2,4,5)cyclophane (10) the synthesis involved a combination of the dithiacyclophane-sulfur extrusion route with a stepwise elaboration of the further bridges. This is illustrated in Scheme 1 [52, 53]. The same scheme has also been utilized to prepare the corresponding tetramethoxy derivative (51), and its analogous bisquinone (52) and quinhydrone (53) [56].

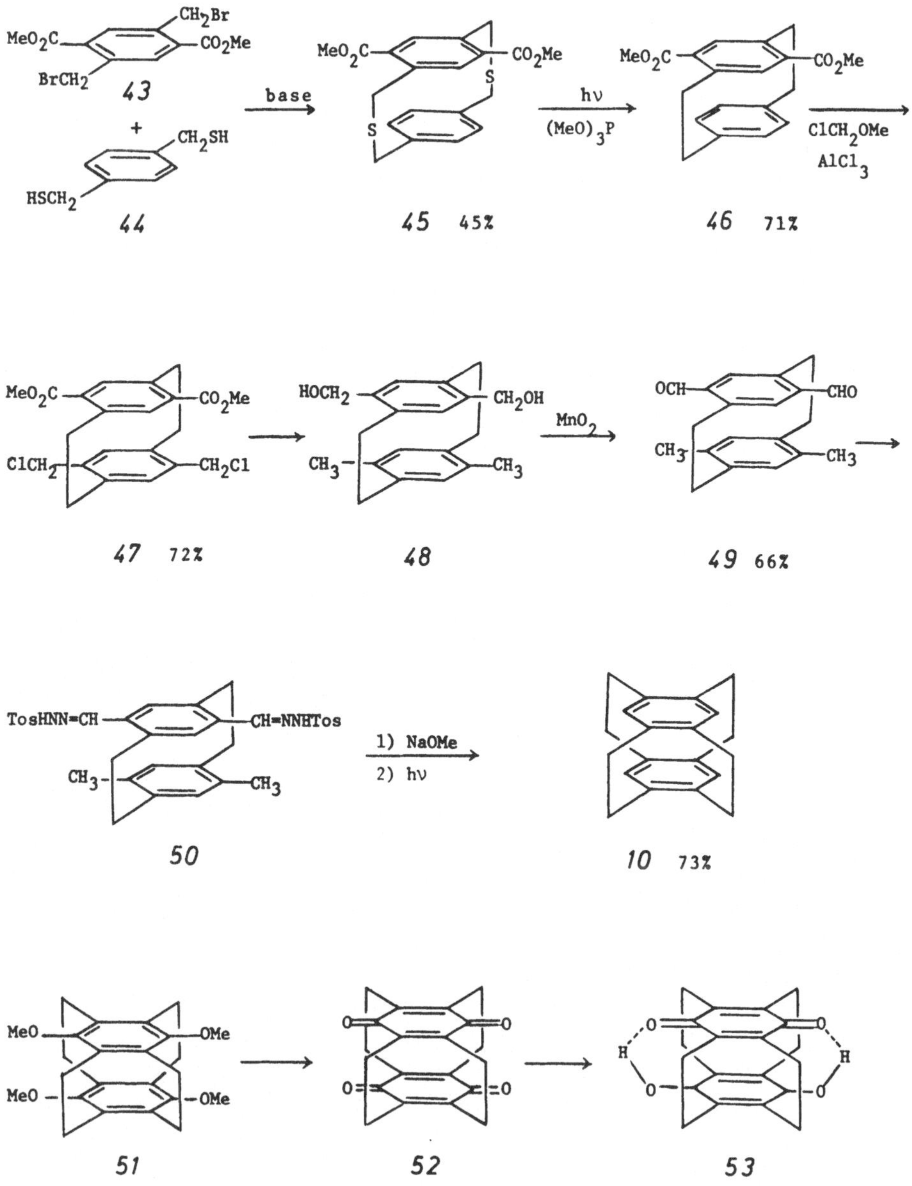

Scheme 1

The elaboration of *46* to provide the two additional bridges illustrates two characteristic properties of [2$_2$](1,4)cyclophane chemistry: (i) the directive effect of the carbomethoxy group leading to pseudo-gem orientation during chloromethylation [6, 57], and ii) the transannular carbene insertion leading to bridge formation and giving (*10*) [6].

2.3 The Diels-Alder Approach

In 1972, Hopf discovered that 1,2,4,5-hexatetraene (54) reacts with acetylenes (55) to give substituted [2₂](1,4)cyclophanes (57) [58-60]. Presumably, the first step is a Diels-Alder reaction to give the p-xylylene (56), which then dimerizes to the [2₂](1,4)cyclophane. In contrast to other cyclophane syntheses involving p-Xylylene intermediates, this synthesis can be carried out in concentrated solutions, giving as much as 60 grams (40–50% yield) of the desired [2₂](1,4)cyclophane per run. This procedure has been standardized and is described in Organic Syntheses [61]. Probably, the reason this reaction is successful is that the rate-controlling step is the Diels-Alder reaction and the concentration of the p-xylylene intermediate (56) never builds up to the point where polymerization or other side reactions dominate.

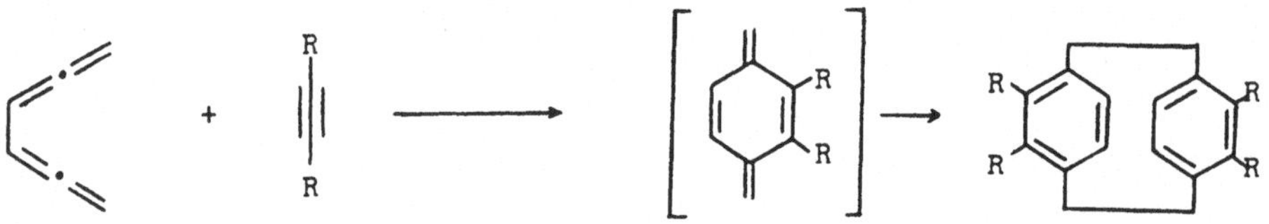

$R = -CO_2Me, -CO_2CMe_3$

$-CN,$ or $-CF_3$

The Diels-Alder addition of methyl propiolate (58) to 1,2,4,5-hexatetraene (54) gives all four possible products: (59), (60), (61), and (62) [62-65]. Although the yield of each individual isomer is low, the method provides them in sufficient quantity that, after separation and purification, each has been used as starting material for constructing a multibridged cyclophane. Thus, as shown in Scheme 2, (59) has been converted to [2₃](1,2,4)cyclophane (6) [62,66]; (60) to [2₄](1,2,3,4)cyclophane (8) [64]; (61) to [2₄](1,2,3,5)cyclophane (9) [63]; and (62) to [2₄](1,2,4,5)cyclophane (10) [65].

The elaboration of the [2₂](1,4)cyclophane derivatives, (59), (60), (61), and (62), to incorporate the additional bridges forming (6), (8), (9), and (10), follows the chemistry worked out by Truesdale and Cram in their synthesis of [2₃](1,2,4)cyclophane (6) [6,67,68], and is analogous to the bridge construction employed in the synthesis of (10) shown in Scheme 1 [52,53]. The use of zinc in dimethylsulfoxide for carbon—carbon bond formation between pseudo-gem bromomethyl substituents is novel, though, and apparently superior to the more commonly used phenyllithium [69].

In addition to serving as starting materials for multibridged cyclophane syntheses, the Diels-Alder adducts of 1,2,4,5-hexatetraene (54) and acetylenes are very useful for preparing various methyl substituted cyclophanes. This is illustrated by the synthesis of 4,5,7,8,12,13,15,16-octamethyl[2₂](1,4)cyclophane (67) [70], shown in

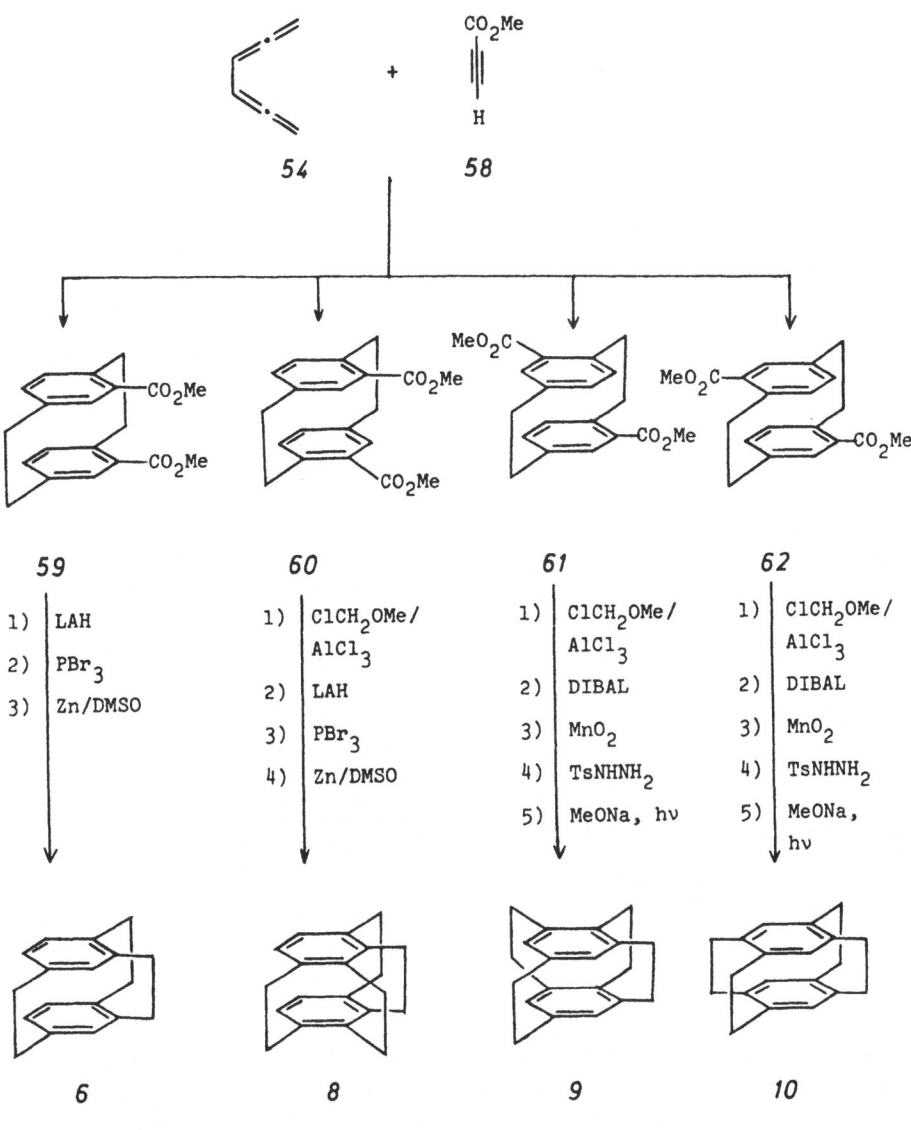

Scheme 2

Scheme 3. 4,5,12,13-Tetramethyl[2₂](1,4)cyclophane (63) is obtained in quantity and in high yield from (57). Treatment of (63) with Cl₂CHOMe and TiCl₄ (Rieche reaction) gives the corresponding aldehyde in excellent yield. Catalytic hydrogenation of the aldehyde gives (64) in 53% yield, but the corresponding acetoxymethyl derivative is isolated as a side-product in 29% yield and this can be recyclized. A four-step sequence of Rieche aldehyde formation, lithium aluminum hydride reduction to the carbinol, reaction with phosphorus tribromide to the bromomethyl derivative, and lithium aluminum hydride reduction gives (65) as a mixture of isomers with the newly-introduced methyl group being present in the isomers at each of the possible positions. A repetition of this four-step sequence then gives the heptamethyl[2₂](1,4)cyclophane (66) in 68% overall yield. Finally, a third repetition of this four-step sequence leads to 4,5,7,8,12,13,15,16-octamethyl[2₂](1,4)cyclophane (67) in 22% overall yield for the final sequence.

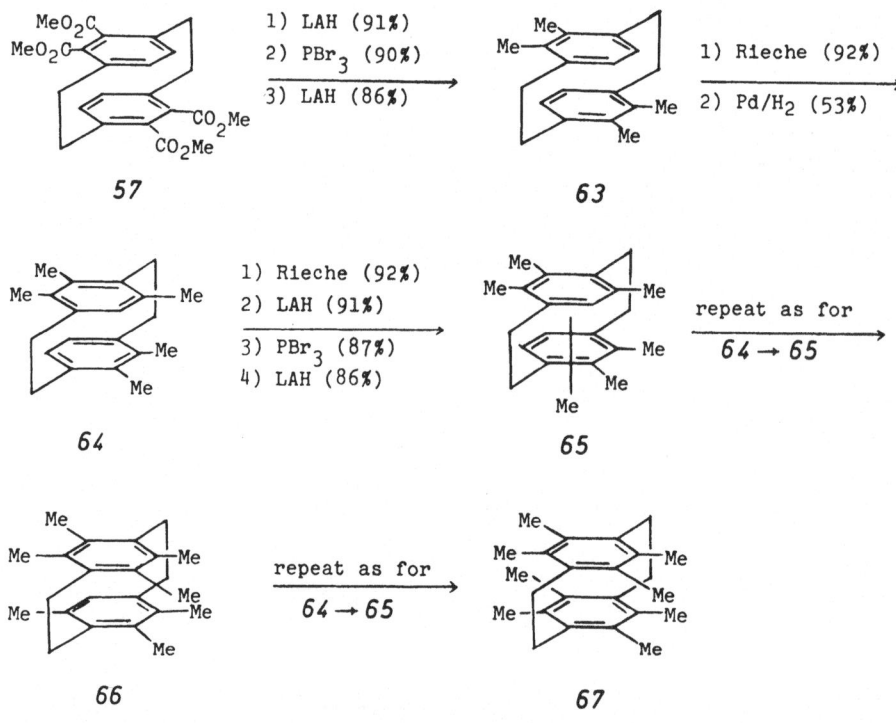

Scheme 3

In a previous report [71], (67) had been described as an extremely labile compound subject to polymerization, presumably due to the high steric strain of the eclipsed methyl groups. However, Eltamany and Hopf found (67) to be extremely stable thermally, being recovered completely unchanged after heating for 17 hours at 230 °C [70].

2.4 The *o*-Xylylene Approach

2.4.1 Gray's Idea

During the course of his experiments on the synthesis of [2₄](1,2,4,5)cyclophane (10), as previously summarized in *Scheme 1*, Richard Gray conceived an idea for preparing (10) in a single step [72]. This idea is outlined in *Scheme 4*. When Gray tried the experiment, subjecting (68) to 700 °C at 10⁻² mm pressure, he found the product to be benzo[1,2;4,5]dicyclobutene (73) and not (10) [48].

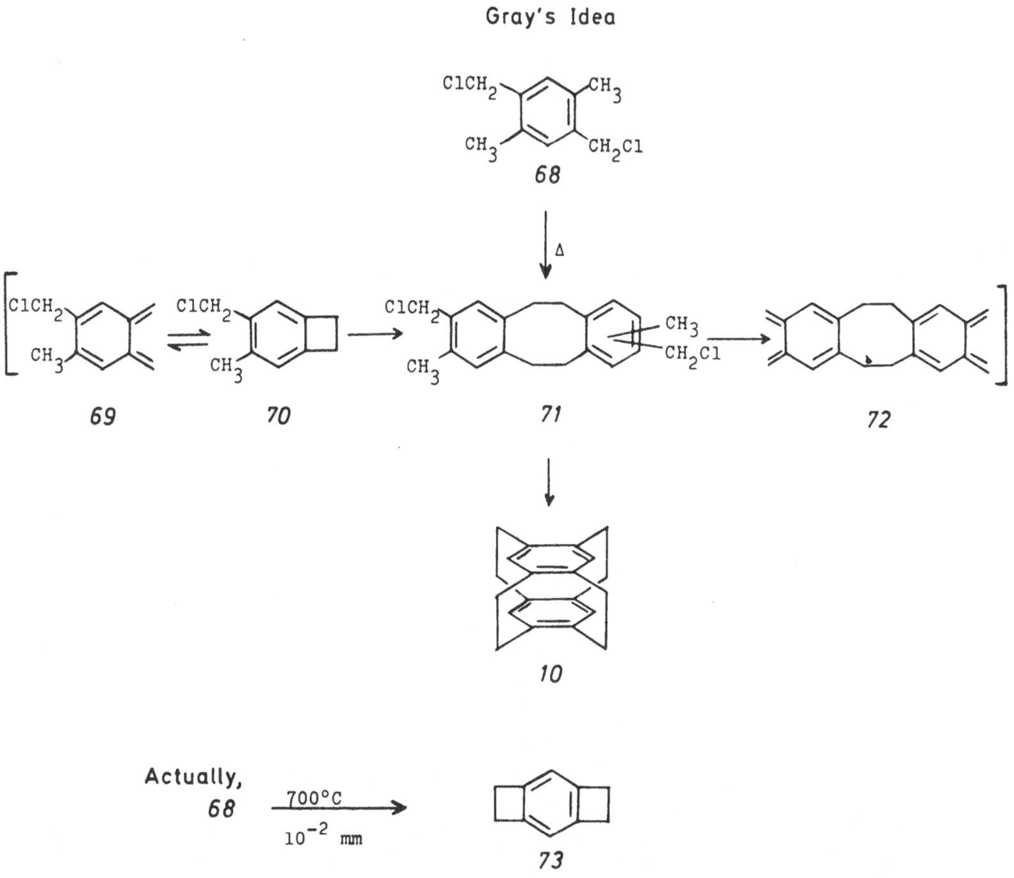

Gray's Idea

ClCH$_2$— —CH$_3$
CH$_3$— —CH$_2$Cl
68

$\downarrow \Delta$

$\left[\begin{array}{} \text{ClCH}_2 \\ \text{CH}_3 \end{array} \right.$ **69** $\rightleftharpoons$ $\begin{array}{} \text{ClCH}_2 \\ \text{CH}_3 \end{array}$ **70** $\longrightarrow$ $\begin{array}{} \text{ClCH}_2 \\ \text{CH}_3 \end{array}$ —CH$_3$ —CH$_2$Cl **71** $\longrightarrow$ $\left. \right]$ **72**

$\downarrow$

10

Actually,

68 $\xrightarrow[10^{-2} \text{ mm}]{700°C}$ **73**

Scheme 4

Later, it was decided to reinvestigate Gray's idea, but do it stepwise. As shown in *Scheme 5*, gas-phase pyrolysis of 2,5-dimethylbenzyl chloride *(74)* gives 4-methyl-benzocyclobutene *(75)* in good yield and in quantity [73, 74]. Its dimerization in solution at 300 °C, following the procedure of Cava [75a], led to *(76)* as a mixture of isomers. Chloromethylation of *(76)* gave both possible isomers, *(77)* and *(78)*, and pyrolysis of this mixture led to the desired $[2_4](1,2,4,5)$cyclophane *(10)* plus *(79)*. Subsequently, it was possible to improve the conditions for this pyrolysis such that *(10)* is the exclusive product in 40% yield [75b].

2.4.2 Syntheses of the Multibridged $[2_n]$Cyclophanes

The primary significance of the synthesis of *(10)*, as shown in *Scheme 5*, was that it showed the feasibility of using intramolecular *o*-xylylene dimerizations for introducing two bridges simultaneously to make multibridged cyclophanes. To extend this method $[2_3](1,2,4)$cyclophane *(6)* was next synthesized starting from 2,4-bis(chloromethyl)toluene *(80)* [76]. Aalbersberg and Vollhardt have also reported a similar

101

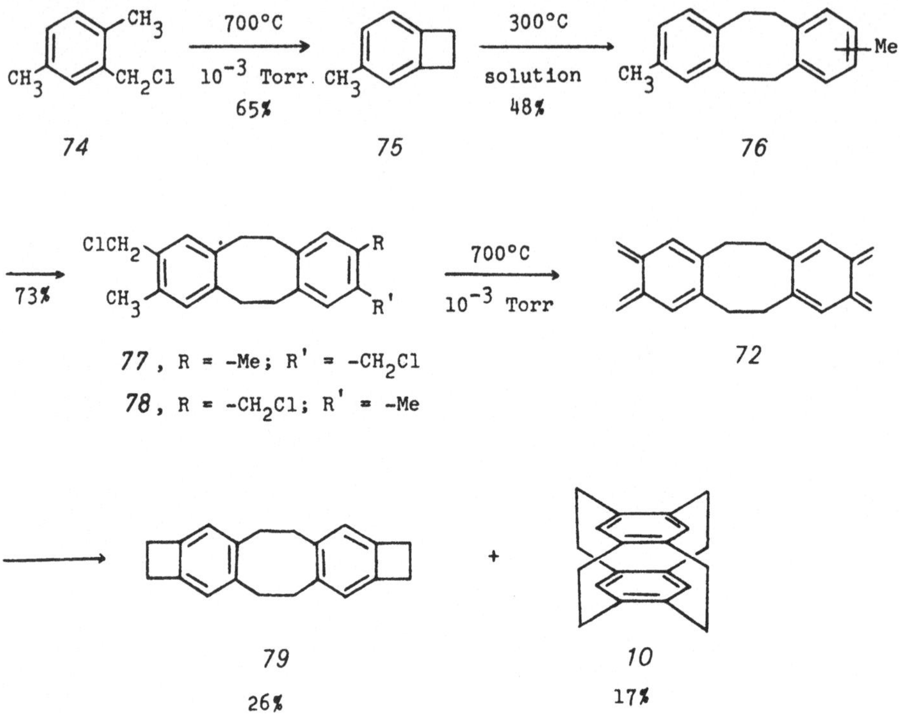

Scheme 5

synthesis of (6) [77]. Further, following the same pattern of reactions, it was possible to convert 2,6-bis(chloromethyl)toluene (81) to (5) and this provided the first synthesis of [2₃](1,2,3)cyclophane (5) [78]. However, 7,14-dimethyl[2₃](1,2,3)cyclophane can readily be obtained by reductive thermolysis of [2₄](1,2,3,5)cyclophane (9), vide infra [79]. These syntheses of (5) and (6) are shown in *Scheme 6*.

With these results in hand, the *o*-xylylene dimerization method was then extended to a synthesis of [2₅](1,2,3,4,5)cyclophane (11), as shown in *Scheme 7* [80,81]. The synthesis of (11) occurs in six steps with an overall yield of 31% and can be carried out on a scale to provide 5 grams of (11) per run. The final pyrolysis is remarkable in that four bridges are created in a single step and in 69% yield.

For the synthesis of the final member of the series, [2₆](1,2,3,4,5,6)cyclophane ((12), superphane) an analogous approach would, in principle, require repeating the *o*-xylylene dimerization three times. In fact, as shown in *Scheme 8*, this was actually done successfully [82,83]. Pyrolysis of 2,4,5-trimethylbenzyl chloride (88) readily gave 4,5-dimethylbenzocyclobutene (89) in quantity. Previously, the dimerization of benzocyclobutenes had been done in boiling diethyl phthalate [75a], but the work-up and isolation of the dimer was very tedious. To avoid this, dimerization in the gas phase was tried using a nitrogen flow at atmospheric pressure to provide concentrations in the hot zone that would increase the number of bimolecular collisions. In this way dimerization of (89) to give (90) occurred cleanly in 63% yield. Thus, intermolecular dimerization of *o*-xylylene intermediates in the gas phase was also found to be a very useful method.

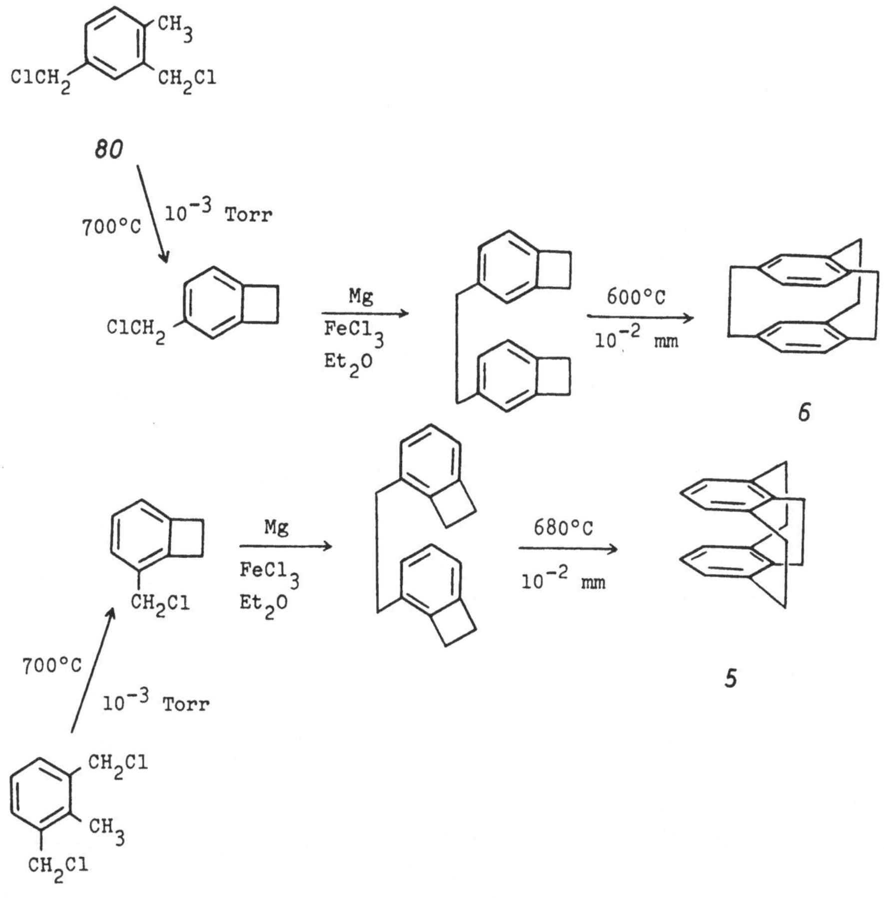

Scheme 6

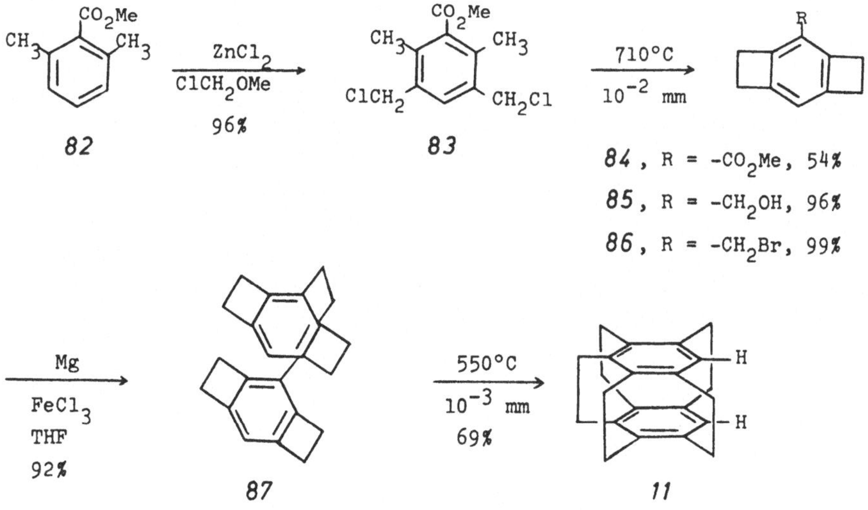

84, R = -CO$_2$Me, 54%

85, R = -CH$_2$OH, 96%

86, R = -CH$_2$Br, 99%

Scheme 7

103

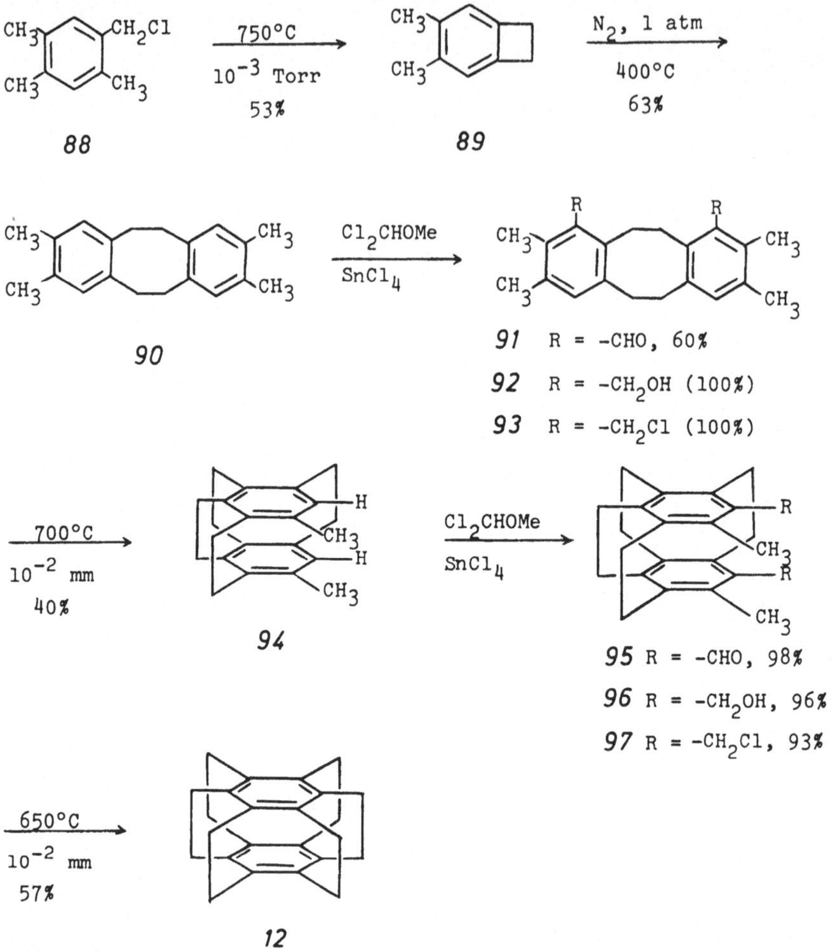

Scheme 8

Subjection of (90) to the four-step cycle of Rieche formylation [84], reduction, conversion to the chloride, and pyrolysis gave 7,14-dimethyl[2$_4$](1,2,3,4)cyclophane (94), and repetition of the four-step cycle with (94) then gave superphane (12). The synthesis of superphane was thus accomplished in ten steps with an overall yield of 4%.

Subsequently, it was found that chloromethylation of [2$_5$](1,2,3,4,5)cyclophane (11) gave the corresponding aldehyde (98) [85]. Conversion of (98) to the tosylhydrazone followed by a carbene insertion reaction occurred smoothly, providing a second, and useful, procedure for preparing superphane [81].

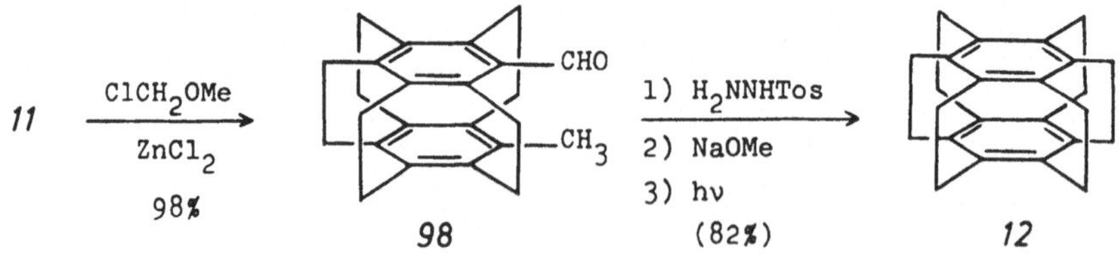

2.5 Miscellaneous Methods

2.5.1 Lewis Acid Catalyzed Rearrangements

In their studies on the chemical properties of [2$_2$](1,4)cyclophane (4), Hefelfinger and Cram observed that, on treatment of (4) with anhydrous hydrogen chloride and aluminum chloride in dichloromethane at -10 °C, rearrangement occurred to give [2$_2$](1,3)(1,4)cyclophane (13) [86, 87]. From a study employing optically active [2$_2$](1,4)-cyclophanes, it was concluded that the rearrangement is intramolecular and probably follows a reaction path involving intermediates such as (99) and (100). The driving force for the rearrangement is though to be the relief in strain in going from (99) to (100), since (4) and (13) are probably fully protonated under the conditions of the reaction [87]. This is an excellent method for preparing (13) itself, although, for preparing [2$_2$](1,3)(1,4)cyclophanes with specific substituents, the dithiacyclophane procedure is superior [54].

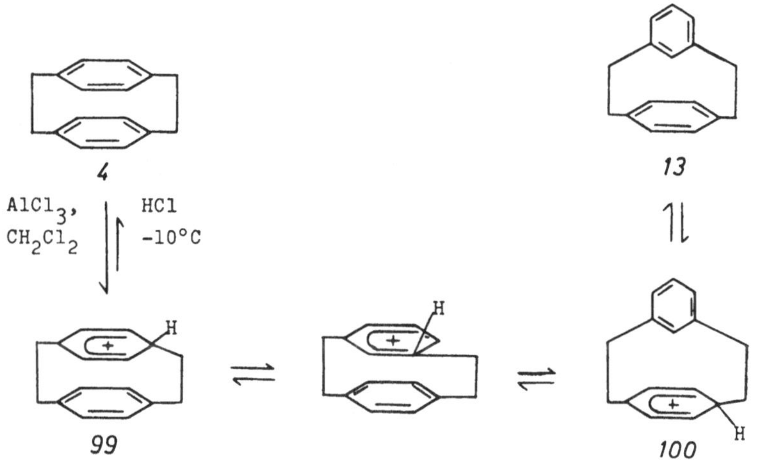

The skewed cyclophanes (14) and (15) are of interest because of their chirality and because their geometry differs from that of the closely-related, symmetrical [2$_n$]-cyclophanes. Nakazaki et al. have prepared (14) and (15) by the dithiacyclophane route shown below [55]. In this case, the yields via the dithiacyclophane route are poor. A more efficient route to [2$_3$](1,2,4)(1,2,5)cyclophane (6) is the Lewis-acid catalyzed rearrangement of [2$_3$](1,2,4)cyclophane (6) [79]. More strikingly, the Lewis-acid catalyzed rearrangement of [2$_3$](1,2,4)cyclophane-9-ene (105) gives [2$_3$](1,2,4)-(1,2,5)cyclophane-9-ene (106) in 85% yield [88]!

2.5.2 Photochemical Approaches

Accompanying their work on the Lewis-acid catalyzed rearrangement of [2$_2$](1,4)-cyclophane (4) to [2$_2$](1,3)(1,4)cyclophane (13), Cram and his colleagues observed that [2$_2$](1,3)(1,4)cyclophane (13) underwent a photochemical rearrangement to give anti-[2$_2$](1,3)cyclophane (3) in 46% yield [89]. Surprisingly, [2$_2$](1,4)cyclophane (4)

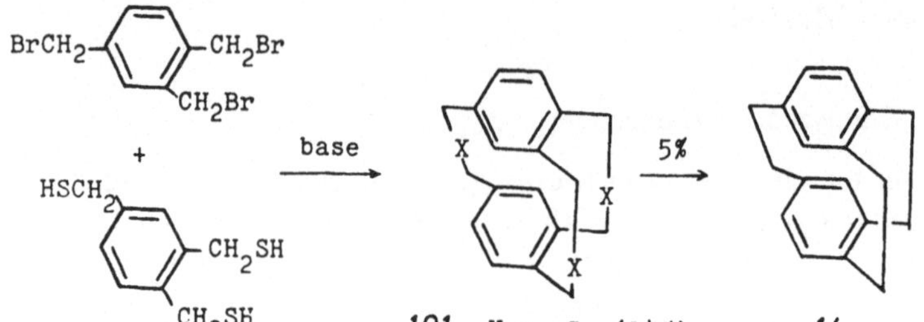

101, X = -S- (14%)   14

102, X = -SO$_2$- (100%)

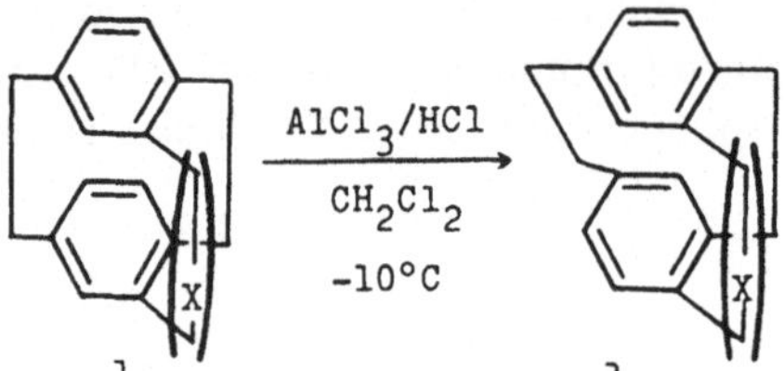

6, ($\frac{1}{3}$X) = -CH$_2$CH$_2$-   15, ($\frac{1}{3}$X) = -CH$_2$CH$_2$-

105, ($\frac{1}{3}$X) = -CH=CH-   (44%)

106, ($\frac{1}{3}$X) = -CH=CH-

(85%)

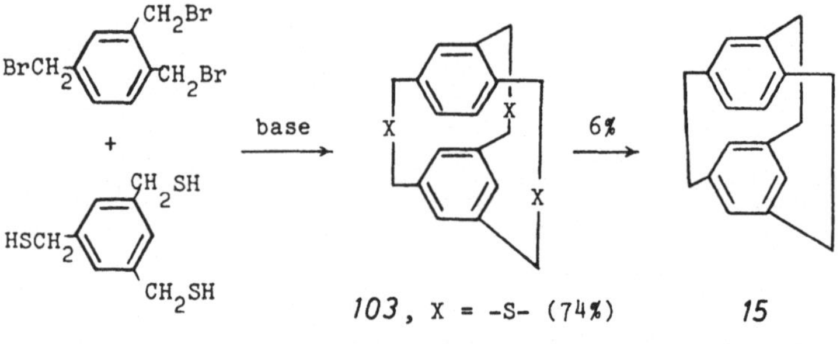

103, X = -S- (74%)   15

104, X = -SO$_2$- (100%)

itself is inert toward irradiation. In an attempt to gain insight regarding this photochemical rearrangement, optically-active (107) was irradiated and found to give an optically-active mixture of three products: (108), (109), and (110). It was suggested that prismane or benzvalene intermediates are probably involved [89].

The photochemical rearrangement of [2$_2$](1,3)(1,4)cyclophane (13) to anti-[2$_2$](1,3)-cyclophane (3) was likewise observed in our studies [90, 91]. However, attempts to apply this photochemical rearrangement to [2$_2$](1,3)(1,4)cyclophane-1,9-dienes (111–113) led only to recovery of starting material and none of the desired anti-[2$_2$](1,3)-cyclophane-1,9-dienes (114–116) could be isolated [90, 91].

A synthesis of the unusual [2$_3$](1,3,5)cyclophane (118) was accomplished by irradiation of the stilbene derivative (117) [92].

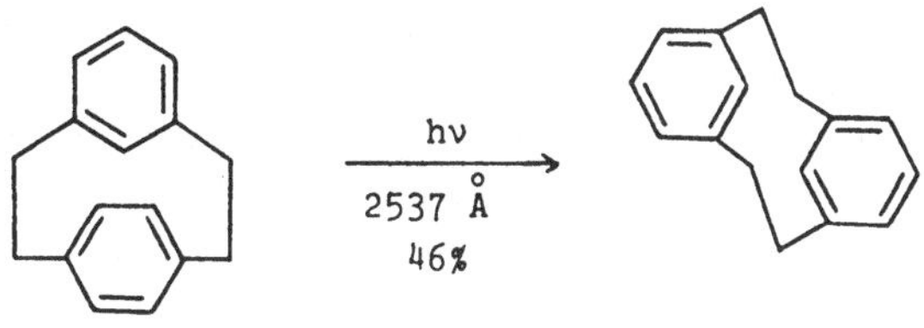

13      $\xrightarrow[\substack{2537\ \overset{\circ}{A}\\46\%}]{h\nu}$ 3

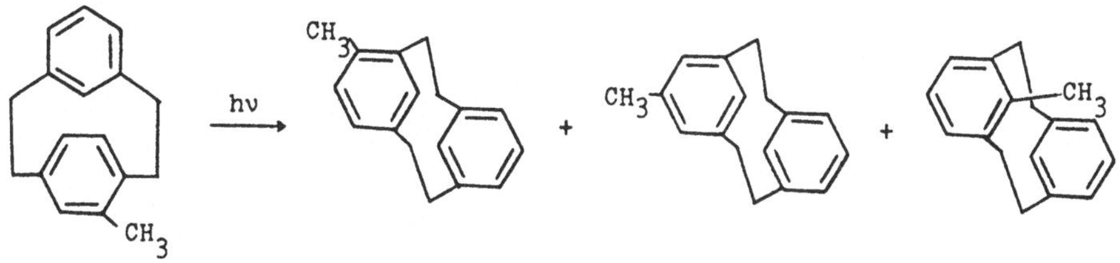

(-)-(R) *107*   *108*   *109*   *110*

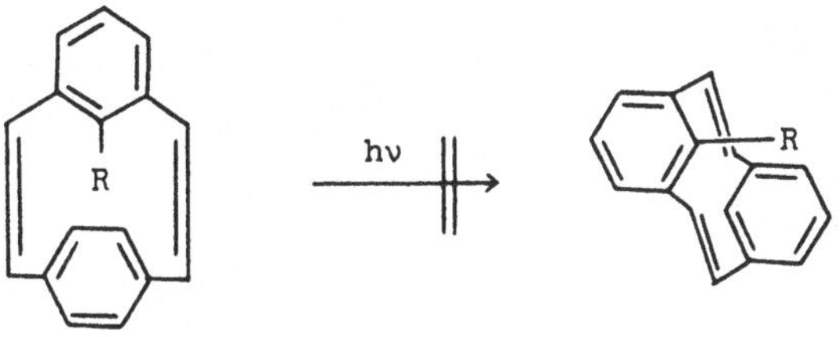

111, R = -H   *114*, R = -H

112, R = -F   *115*, R = -F

113, R = -CH$_3$ *116*, R = -CH$_3$

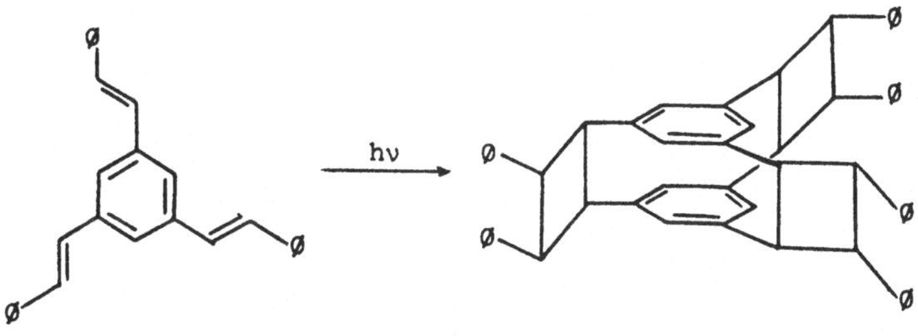

117   *118*

2.5.3 Bis(dithiane) Alkylation

Seebach, Jones, and Corey first introduced the alkylation of bis(dithianes) as a method for converting aldehydes to cyclic diketones [93]. In an extension of this method it was found that isophthaladehyde bis(dithioacetal) (119) could be alkylated with either *m*-xylylene dibromide or *p*-xylylene dibromide to give the corresponding bridge-substituted cyclophanes (120) and (121) [90, 94]. These, in turn, can readily be

107

desulfurized with Raney nickel to provide the parent cyclophanes (3) and (13), or hydrolyzed in the presence of mercuric chloride to give the corresponding cyclophane-1,9-diketones (122) and (123). Gschwend has employed this approach to prepare the optically active 1-hydroxy- and 1-keto-*anti*-[2₂](1,3)cyclophanes for measuring the energy barrier to conformational flipping in the *anti*-[2₂](1,3)cyclophane series (vide infra, p. 27) [95].

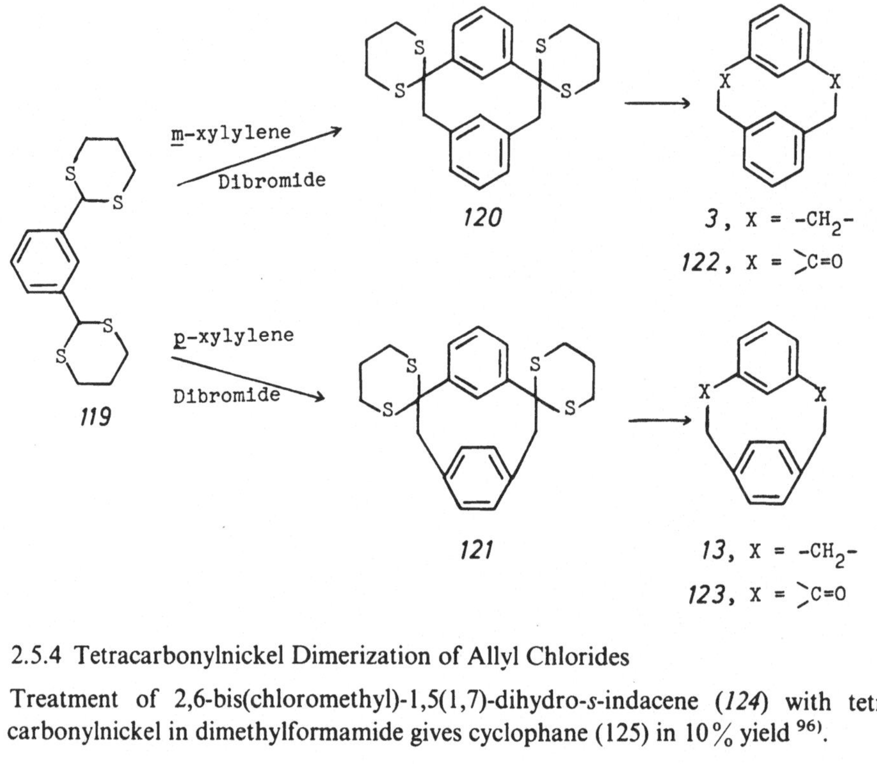

120

3, X = −CH₂−

122, X = >C=O

119

121

13, X = −CH₂−

123, X = >C=O

2.5.4 Tetracarbonylnickel Dimerization of Allyl Chlorides

Treatment of 2,6-bis(chloromethyl)-1,5(1,7)-dihydro-*s*-indacene (*124*) with tetra-carbonylnickel in dimethylformamide gives cyclophane (125) in 10% yield [96].

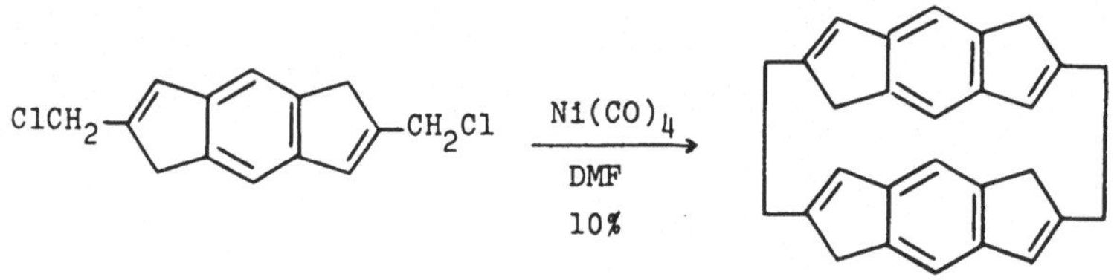

124

125

3 Properties of [2ₙ]Cyclophanes

3.1 Introductory Remarks and Geometry

The outstanding characteristic of the [2ₙ]cyclophanes is the interaction of the π-electron systems of the two benzene decks, combining to give a new, overall π-electron system whose highest occupied molecular orbital (HOMO) is of higher energy than that of a corresponding alkyl benzene and whose lowest unoccupied molecular orbital (LUMO) is lower in energy than that of a corresponding alkyl benzene. This doubling of the molecular orbitals and the narrowing of the HOMO-LUMO gap, as compared to benzene, directly relates to many of the differences in chemical and physical properties of the [2$_n$]cyclophanes as compared to benzene. The same type of interaction, although weaker, is still present in [3$_n$]cyclophanes but is absent in the [4$_n$]cyclophanes, where the individual decks behave as separate, isolated π-electron systems [97, 98]. A molecular orbital analysis of the [2$_n$]cyclophanes is presented in the accompanying chapters in this volume on the photoelectron spectra [99] and esr spectra [100] of the [2$_n$]cyclophanes, so further discussion of this aspect will be limited here to its relevance regarding specific physical and chemical properties.

The second important characteristic of the [2$_n$]cyclophanes is the forced deformation of the benzene rings and the relationship of their geometry to their properties. Single crystal x-ray structural analyses have been made of the following [2$_n$]cyclo-phanes: *anti*[2$_2$](1,3)cyclophane (3) [101, 102] and its various derivatives [103–106], [2$_2$]-(1,4)cyclophane (4) [3, 107, 108] and its various derivatives [108, 109], [2$_3$](1,2,3)cyclophane (5) [110], [2$_3$](1,3,5)cyclophane (7) [111] and its corresponding triene [112], 4,13-dimethyl-[2$_4$](1,2,3,4)cyclophane [110], [2$_4$](1,2,4,5)cyclophane (10) and its Birch reduction product [113], [2$_5$](1,2,3,4,5)cyclophane [110], and [2$_6$](1,2,3,4,5,6)cyclophane [110].

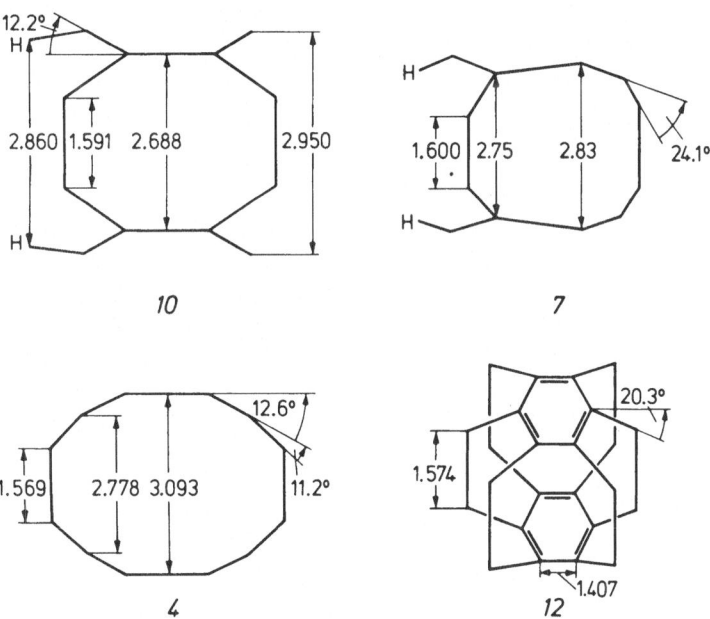

Fig. 1. Side profile views of [2$_2$](1,4)cyclophane (4) [108], [2$_3$](1,3,5)cyclophane (7) [111], and [2$_4$](1,2,4,5)-cyclophane (10) [113], and a full view of superphane (12) [110], as determined by single crystal, x-ray analysis

Misumi et al. have shown, in their studies of the conformational equilibria of multilayered [2₂](1,3)cyclophanes [13, 16, 114, 115], that a boat deformation of a benzene ring is more stable than a chair deformation by at least 4 kcal/mol [115, 116]. Presumably, deformation of a benzene ring to a boat conformation retains better π-orbital overlap than does deformation to a chair. Thus, it is not surprising that boat deformation of the benzene rings is a common occurrence among the [2$_n$]cyclophanes.

Although boat deformation of the benzene rings occurs with [2₂](1,3)cyclophane (3), [2₂](1,4)cyclophane (4), [2₄](1,2,4,5)cyclophane (10), [2₅](1,2,3,4,5)cyclophane (11), and [2₂](1,3)(1,4)cyclophane (13) [87], the extent of the boat deformation is most striking for (4) and (10), as shown in Figure 1. [2₃](1,2,3)Cyclophane (5) and [2₄](1,2,3,4)cyclophane (8) have open butterfly-type structures, whereas [2₃](1,3,5)-cyclophane (7) shows a small chair deformation. Only in the case of the most highly-strained and highly-bridged member, superphane (12), do the benzene rings remain planar (see Fig. 1).

Among the [2$_n$]cyclophanes the mean planar distance between benzene decks varies from 2.99 Å for [2₂](1,4)cyclophane (4) to 2.62 Å for superphane (12). Substituents attached to the benzene ring, whether hydrogen atoms or larger, do not lie in the plane of the ring but are distorted toward the center of the cyclophane molecule. An obvious explanation for this bond angle deformation would be that there is a rehybridization of the aromatic carbons giving them more sp^3 character. Such a rehybridization would relieve electron density in the congested region between decks and increase the electron density outside of the rings. The $^{13}C-H$ coupling constants are commonly used as a measure of the s and p character of a carbon-hydrogen bond, and the observed $^{13}C-H$ coupling values for the [2$_n$]cyclophanes are in the normal range for sp^2 hybridization [51]. Even so, the measured $^{13}C-H$ coupling constants may be the result of a combination of effects and not a true measure of the s and p character of the aromatic carbons present in [2$_n$]cyclophanes.

3.2 Raman Spectra

A laser Raman spectral study has been made of the [2$_n$]cyclophanes [117]. The in-plane force constants for stretching or bending of the aromatic rings of the individual [2$_n$]cyclophanes are nearly unchanged from those of the corresponding methylated benzenes. Although strong mechanical coupling of the two decks may exist, there are no obvious correlations in the spectra that could be attributed to a cyclophane vibrational mode. The CH₂ and CH bands often appear at higher frequencies than expected, probably due to steric effects.

3.3 Ultraviolet Spectra

As would be expected from the molecular orbital analysis of the [2$_n$]cyclophanes, their ultraviolet spectra are quite different from those of the corresponding alkylbenzenes. A number of theoretical and empirical attempts have been made to correlate and interpret the electronic spectra of [2$_n$]cyclophanes [118-123]. As the prototype for the series, [2₂](1,4)cyclophane (4) has absorption bands at 225 nm

(ε, 25200), 244 (3,163, sh), 286 (324), and 302 (199) [119]. The absorption band at 302 nm is well beyond the long wavelength absorption of simple alkyl benzenes. It has been termed the "cyclophane band" and is characteristic of the [2_n]cyclophanes.

Probably the two most important factors governing the position and intensity of the "cyclophane bands" are 1) the mean planar distance between decks, and 2) the extent and nature of the deformation of the benzene rings. In Table 1 the long wavelength absorption bands of the [2_n]cyclophanes are summarized.

Table 1

[2_n]Cyclophane	Structure	$\lambda_{max}(\varepsilon)$		Ref.
[2_2](1,4)-	(4)	286 nm (324)	302 nm (199)	119)
[2_2](1,3)-	(3)	272 (436)	277 (356)	99,127)
[2_3](1,3,5)-	(7)	258 (1200)	312 (96)	51)
[2_3](1,2,4)-	(6)	290 (393)	309 (180)	66)
[2_3](1,2,3)-	(5)	272 (389)	275 (379)	78)
[2_4](1,2,4,5)-	(10)	294 (660)	303 (1050)	53)
[2_4](1,2,3,5)-	(9)	287 (370)	308 (195)	63)
[2_4](1,2,3,4)-	(8)	283 (370)	297 (280)	64)
[2_5](1,2,3,4,5)-	(11)	294 (352)	313 (200)	81)
[2_6](1,2,3,4,5,6)-	(12)	296 (421)	311 (324)	83)
[2_2](1,3)(1,4)-	(13)	284 (280)	292 (230)	86)
[2_3](1,2,4)(1,2,5)-	(14)	285.5 (640)	293.5 (530)	55)
[2_3](1,2,4)(1,3,5)-	(15)	235 (10,030)	300 (380)	55)
	(126)		280 (28,000)	33)
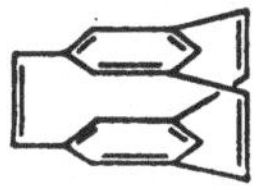	(127)	252 (1960)	325 (90)	51)
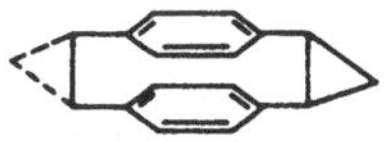	(128)	260 (5000)	304 (900)	125)

Superphane (12), which has the shortest distance between decks (2.624 Å), but completely planar benzene rings, has its "cyclophane band" at 311 nm, whereas [2_3](1,3,5)cyclophane (7) and [2_5](1,2,3,4,5)cyclophane (11), which have a greater distance between decks but have distorted benzene rings, have their "cyclophane bands" at 312 and 313 nm, respectively.

The effect of unsaturation in the bridges is not consistent. [2_2](1,4)Cyclophane-1-ene and [2_2](1,4)cyclophane-1,9-diene have essentially the same ultraviolet absorption spectrum as [2_2](1,4)cyclophane (4) itself [124]. However, anti-[2_2](1,3)cyclophane-1,9-diene (126), which exhibits only a small shift to longer wavelength, as compared to anti-[2_2](1,3)cyclophane (3), shows a dramatic increase in intensity. On the other hand, [2_3](1,3,5)cyclophane-1,9,17-triene (127) shows a long wavelength shift of 13 nm as compared to (7), but virtually no change in intensity. The "cyclophane band" of (127) has thenlongest wavelength band of any unsubstituted $[2_n]$cyclophane.

Gleiter et al. have examined the photoelectron spectrum of (128), the cyclopropano analog of (4), and conclude that there is a strong $\sigma-\pi$ interaction between the cyclopropane bridges and the benzene decks [125, 126]. The long wavelength band of (128) is little changed from that of (4) and the only evidence in the electronic spectrum of (128) to suggest an unusual $\sigma-\pi$ interaction is the strong increase in intensity of the "cyclophane band".

3.4 NMR Spectra and Conformational Analysis

Proton nuclear magnetic resonance has been invaluable in providing information regarding the structure and geometry of $[2_n]$cyclophanes. The $[2_n]$cyclophanes can be divided into three main classes: 1) those in which the benzene decks are directly over each other, as in (4). 2) those of fixed geometry where there is only partial overlap of the benzene decks, as in (3). and 3) those having partial overlap of the benzene decks and showing conformational mobility, as in (13).

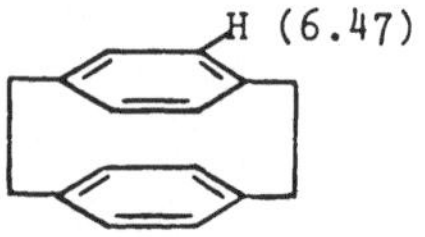

4   H (6.47)

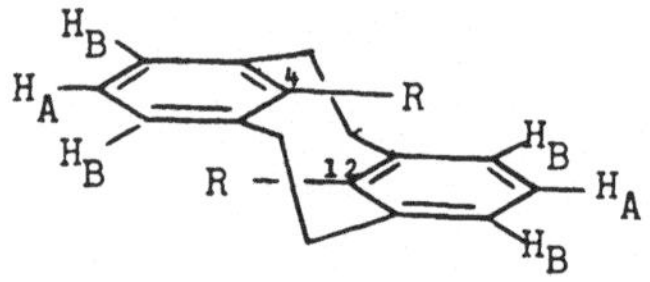

3 , R = -H (4.25)
3a , R = -Me (0.56)
AB$_2$ multiplet (6.98-7.30)

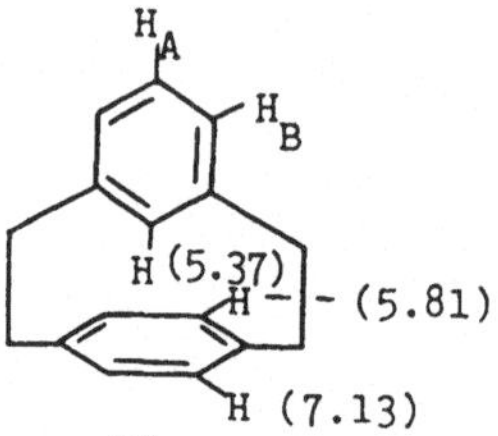

13
A$_2$B m (6.6-7.10)
H (5.37) - (5.81) 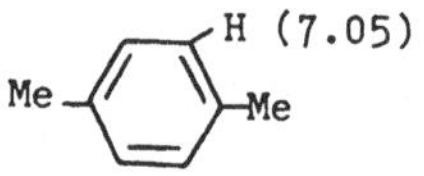
H (7.13)

129   H (7.05)

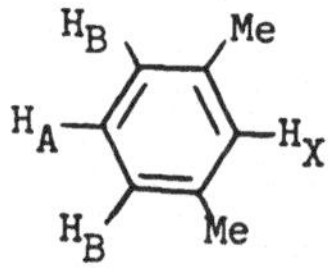

130
AB$_2$X multiplet (6.92-7.22)

In [2$_2$](1,4)cyclophane (4) the signal for the aromatic proton H_A appears at δ 6.47, shifted about 0.6 ppm upfield from that of p-xylene (129) [128]. This upfield shift of the aromatic protons of a cyclophane is regarded as being due to a shielding effect from the ring current of the benzene ring in the opposite deck. For the other $[2_n]$cyclophanes, where the benzene rings are directly over each other, similar upfield shifts are observed, varying in extent dependent on the distance between decks and changes in geometry of the benzene rings.

In the case of *anti*-[2₂](1,3)cyclophane (3), the H_A and H_B aromatic protons appear as an AB_2X multiplet a δ 6.98–7.30 [129], shifted only slightly from the signal of the AB_2X multiplet of *m*-xylene itself. Thus, there is very little shielding of these protons by the ring current of the benzene ring in the opposite deck. However, the internal aromatic proton (R in (3)) is directly over the center of the opposite benzene ring and is very strongly shifted upfield (Δδ = 2.75 ppm), appearing at δ = 4.25 [121, 127, 129–133]. For the case of (3a), the protons of the internal methyl groups appear at δ 0.56, showing again a strong upfield shift of 1.6 ppm from those of the methyls in 1,2,3-trimethylbenzene [129]. The difference between the upfield shifts of the internal protons of (3) and (3a) arises from two effects. The geometry of (3a) differs from that of (3), having a C-4 to C-12 distance of 2.79 Å compared to 2.66 Å for (3) [101, 103]. Secondly, the protons on the methyl group are two bonds removed from the aromatic ring and are not so centrally located over the opposite benzene ring.

The energy barrier to conformational flipping of the *anti*-[2₂](1,3)cyclophanes is sufficiently high so that the molecules are commonly regarded as being rigid. However, in a remarkable study, Gschwend has measured the rates of racemization of the optically-active 1-hydroxy-*anti*-[2₂](1,3)cyclophane (131) and its corresponding ketone (131a) [110]. The $\Delta G^{\ddagger}_{150}$ value for the alcohol (131) is 33 kcal/mol and the $K_{equilibrium}$, $K_{equatorial}$-OH/K_{axial}—OH = 1.49. For the ketone (131a), $\Delta G^{\ddagger}_{150} = 25$ kcal/mol. In many respects, it is surprising that the energy barrier for these isomerizations are as low as they are. In the thermal racemization of optically-active [2₂](1,4)cyclophane derivatives, vide infra [57], $\Delta G^{\ddagger}_{200} = 38$ kcal/mol, and bond breaking occurs to give intermediate diradicals. Gschwend was unable to detect ot trap radical intermediates during the racemization of the *anti*-[2.2](1,3)cyclophanes. Also, the chiral atom bearing hydroxyl in (131) showed no racemization, whereas the presence of radical intermediates should have caused some racemization of this chiral atom. Gschwend concludes that the reversible isomerization between (131) and (132) does not involve a bond breaking process.

Although the entropy term for the racemization of the alcohol (131) is quite small, the ketone (131a) has an amazingly large, and unexplained, entropy term (−48.4 ± 0.6 e.u.), being the major factor in the $\Delta G^{\ddagger}$ value for the ketone. The exact mechanism for the isomerization of *anti*-[2.2](1,3)cyclophanes remains unknown.

There is no evidence that the thermal isomerization of *anti*-[2₂](1,3)cyclophanes (3) involves *syn*-[2₂](1,3)cyclophanes (2) as intermediates. The parent substance, *syn*-

131, R = –OH; R' = –H   *132*, R = –OH; R' = –H

131a, R & R' = $\overset{O}{\underset{\|}{}}$ *132a*, R & R = $\overset{O}{\underset{\|}{}}$

113

[2₂](1,3)cyclophane (2), is still unknown, although derivatives, as shown by structures (133a–c) have been prepared and characterized [134]. Their ^1H NMR spectra are in accord with those of the other [2ₙ]cyclophanes where the benzene decks are directly over each other. Thus, the protons of the methyl groups of (133a–c) appear in the region of δ 2.10–2.20, as is normal for toluene and xylene methyls. The syn-[2₂](1,3)-cyclophanes (133a–c) readily isomerize thermally and are converted on melting to the corresponding anti-[2₂](1,3)cyclophanes (134a–c) in high yield [134].

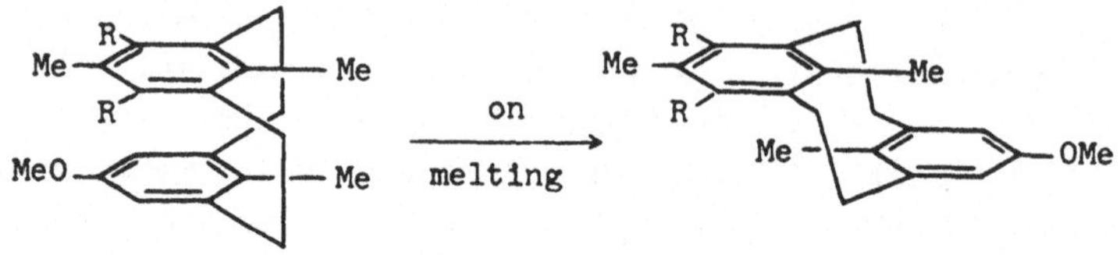

133a, R = –CN	134a, R = –CN
133b, R = –CHO	134b, R = –CHO
133c, R = –CH₂OH	134c, R = –CH₂OH

The example for the third category of conformationally mobile [2ₙ]cyclophanes is [2₂](1,3)(1,4)cyclophane (13). At room temperature, the ^1H NMR spectrum of 13 is in accord with a geometry in which the meta-bridged ring is partially overlapping and more or less parallel to the para-bridged ring. Thus, the outside aromatic protons of the para-bridged ring are in the usual aromatic region (δ 7.13), whereas the inside aromatic protons feel the ring current of the opposite deck and are shifted upfield to δ 5.81 [87]. The meta-birdged ring overlaps the opposite deck more than in the case of the anti-[2₂](1,3)cyclophane, and the AB₂ multiplet shows a small but definite upfield shift (Δδ ≅ 0.3 ppm) due to the ring current from the opposite deck. The interior proton of the meta-bridged ring again shows a large ring current effect, appearing at δ 5.37. When the temperature of the probe of a 100 MHz instrument is raised to 157 °C, coalescence occurs and a time-averaged spectrum results [86, 135–137]. The kinetic parameters for the conformational flipping process for 13 have been determined to be ΔH⁺ = 17.0 ± 0.5 kcal/mol., ES⁺ = −8.8 ± 2.4 e.u.,

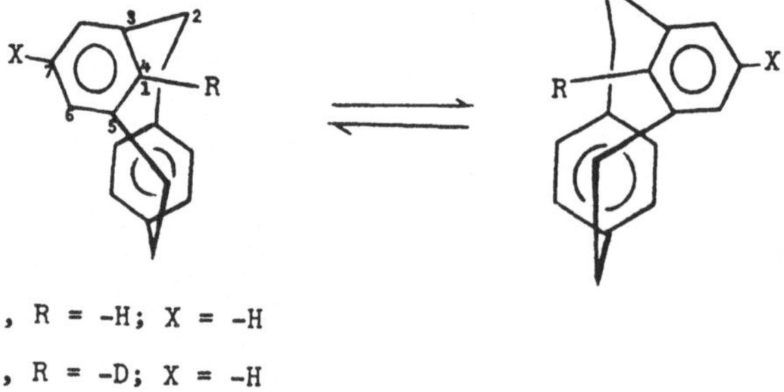

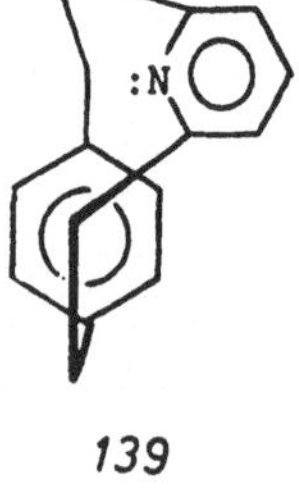

135, R = –H; X = –H
136, R = –D; X = –H
137, R = –H; X = –NH₂
138, R = –H; X = –NO₂

139

and $\Delta G^+ = 20.6$ kcal/mol. [135]. Presumably, the transition state for conformational flipping has the meta-bridged ring perpendicular to the para-bridged ring with the interior proton of the meta-bridged ring thrust into the cavity of the π-electron cloud of the para-bridged ring and being only about 2.1 Å from the aromatic carbon framework [135].

Severe crowding of the C(4)-hydrogen in the transition state would be predicted to cause marked changes in the vibrational frequencies of the C(4)-H bond, detectable in an isotope effect that appropriately is termed a steric isotope effect in view of the source of the vibrational changes. When 4-deuterio-[2$_2$](1,3)(1,4)cyclophane (136) was prepared and its rate of conformational flipping measured, the k_D/k_H ratio was found to be 1.20 ± 0.04, the largest steric deuterium isotope effect thus far observed [135]. Similarly, remote substituents placed at carbon-7 have a marked effect. Thus, the rate of conformational flipping of the 7-amino derivative (137) is only 0.189 times as fast as for the parent compound [135]. Surprisingly, the rate of conformational flipping for the 7-nitro derivative (138) is likewise slower, being only 0.705 that of (135) [135]. As a further comparison, 4-aza-[2$_2$](1,3)(1,4)cyclophane (139) was prepared [138]. In this case coalescence of the ^{1}H NMR spectrum occurs at -43.5 °C, corresponding to a $\Delta G^+ = 10.7$ kcal/mol. Again, this demonstrates the smaller steric factor of a nitrogen lone pair as compared to a carbon-hydrogen bond.

It should also be noted that the conformational flipping of cyclophanes having meta-bridged rings, other than six-membered joined to para-bridged benzenes, has also been studied. Thus, the azulenophanes (140) and (141) have been made and studied. Conformational flipping for the five-membered ring, as in (140), is more difficult, $\Delta G^+ \geq 22.0$ kcal/mol. [139] than for the six-membered ring in (13). This is as would be expected since the smaller internal bond angles of the five-membered ring require greater penetration into the π-electron cavity by the interior carbon-hydrogen bond in the transition state. Also, the azulene polarization increases the electron density of the interior σ carbon-hydrogen bond.

In contrast, conformational flipping for the seven-membered ring, as in (141), is very much easier, $\Delta G^+ = 16.2$ kcal/mol. [140,141]. In this case the greater internal bond angles of the seven-membered ring require a much shallower penetration by the interior carbon-hydrogen bond into the π-electron cavity in the transition state. Furthermore, the azulene polarization leads to a decrease in electron density of the σ carbon-hydrogen bonds in the seven-membered ring.

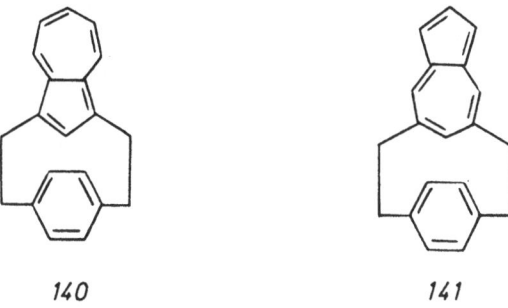

140   141

3.5 [2ₙ]Cyclophanes with Unsaturated Bridges

The conversion of [2ₙ]cyclophanes to derivatives having unsaturated bridges was first demonstrated by Dewhirst and Cram [124]. Bromination of [2ₙ](1,4)cyclophane (4) with N-bromosuccinimide followed by dehydrohalogenation of the product gave [2ₙ](1,4)cyclophane-1,9-diene (42). However, with the introduction of the dithiacyclophane route, ring contraction of the dithiacyclophanes by Stevens [33], Stevens-benzyne [42], or Wittig rearrangements [39], followed by Hofmann eliminations [33] or sulfoxide pyrolyses [42] provided cyclophanes with unsaturated bridges in a more convenient fashion. As discussed previously, (40) is readily prepared from 2,11-dithia-[3₂](1,4)cyclophane (42) [42]. It is somewhat surprising that the ^{1}H NMR signal for vinyl protons in cyclophane-dienes, where the benzene rings are directly over each other, appear at lower field ($\delta = 7.20$) than do the aromatic protons ($\delta = 6.50$) [42,51].

142

143, R = –H

144, R = –Me

145, R = –H

146, R = –Me

Similarly, 2,11-dithia[3₂](1,3)cyclophanes (142) are converted in high yield to *anti*-[2₂](1,3)cyclophane-1,9-dienes [33]. When alkyl groups are present at C-4 and C-12 positions, as in (144), spontaneous valence tautomerization occurs giving the corresponding dihydropyrene derivative (146) [33,142]. In the case of (143) itself, isolation and characterization is possible [33], including a single crystal x-ray analysis [105]. In comparison to *anti*-[2₂](1,3)cyclophane (3), where the internal protons are observed at δ 4.25, the internal protons at C-4 and C-12 of (143) appear at δ 7.90. This large downfield shift may, in part, be due to some deshielding exerted by the bridging vinyl groups, but the major influence is more likely the general flattening of the cyclophanediene moving the internal protons from a shielding region to a deshielding region with respect to the benzene ring of the opposite deck.

The deep green dihydropyrenes (145 and 146) are examples of bridged [14]annulenes having a rigid, planar perimeter of equivalent carbon-carbon bonds [144,145]. The signal for the internal protons of (145) appear at δ-5.49 and that of the internal methyl protons of (146) at δ-4.25, showing the exceptionally strong ring current present in these bridged [14]annulenes [33]. Measurements of the thermal equilibrium between (144) and (145) have shown the dihydropyrene structure (145) to be of lower energy, but only by 2.5 kcal/mole [145]. The interconversion of (144) and (145) can also be accomplished photochemically [146], and this photochemical behavior has been examined in detail for about thirty derivatives of dihydropyrene [145]. A theoretical interpretation of this photochemical behavior has also been made [147].

A Stevens rearrangement followed by a Hofmann elimination was also successful in converting 2,11-dithia-[3₂](1,3)(1,4)cyclophane (148) to [2₂](1,3)(1,4)cyclophane-1,9-diene (149 [90]). Although the ^{1}H NMR spectrum of (149) is temperature dependent, indicating conformational flipping, its coalescence temperature is —96 °C, which corresponds to $\Delta G^\ast = 8.3$ kcal/mol. The time-averaged, room-temperature, spectrum of (149) shows the aromatic protons of the para-bridged ring as a singlet at δ 6.81 and the C-4 proton at δ 4.29. The sharply-lowered energy barrier for the conformational flipping of (149), as compared to [2₂](1,3)(1,4)cyclophane (13), is due to two factors: 1) the changed geometry doesn't require as deep a penetration of the C-8 proton into the π-electron cavity and 2) the energy of the transition state is lowered by conjugative stabilization as the meta-bridged aromatic ring becomes coplanar with the vinyl bridges.

$$
148 \quad \xrightarrow[\text{2) Hofmann}]{\text{1) Stevens}} \quad 149 \qquad 150
$$

148   149   150

149: H (δ4.29), H (δ6.81)

150: H (δ6.33), H (δ7.39), H (δ6.86)

The corresponding 4-aza[2₂](1,3)(1,4)cyclophane-1,9-diene (150) was also prepared and its ^{1}H NMR spectrum was symmetrical and unchanged by cooling to the lowest temperature (—110 °C) experimentally feasible [138]. An x-ray analysis of (150) showed the two aromatic rings to be perpendicular to each other [148], and presumably this geometry is present also in solution. For (150), the combination of reduced steric interaction and conjugative stabilization lowers the energy of the perpendicular orientation of the aromatic rings to such a degree that this geometry is now that of the ground state of the molecule.

As yet none of the more highly-bridged [2_n]cyclophanes has been made with only vinyl groups as bridging members. Even the question of what sort of interaction adjacent vinyl groups bridging a cyclophane will have remains to be settled. A strong stimulus for synthesizing [2₆](1,2,3,4,5,6)cyclophanehexaene (151) is the possibility of obtaining a circularly, delocalized π-electron system involving the π-electrons of the vinyl bridges.

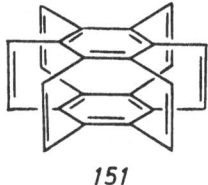

151

Another area for exploration is the synthesis of [2_n]cyclophanes having acetylenic bridges. Psiorz and Hopf have shown that the dehydrohalogenation of (152) using *tert*-butyllithium in tetrahydrofuran leads to the highly symmetrical trimer (154) [149]. It is assumed that the acetylenic-cyclophane (153) is an intermediate.

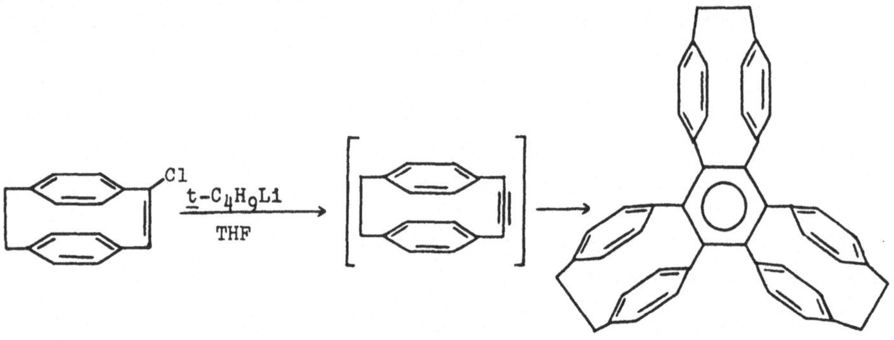

152   153   154

3.6 Chemical Properties

3.6.1 Ring Strain and Radical Cleavage of the Ethano Bridges

The bond angle and bond length deformations, as well as the face to face compression of two benzene rings, leads to ring strain in the [2$_n$]cyclophanes. For cyclophanes (3), (4), and (13), Boyd has measured their heats of combustion and concluded that their strain energies are 12, 31, and 23 kcal/mol, respectively [150a, b; 151]. Similar studies have not been done for the other [2$_n$]cyclophanes. However, Lindner has calculated the strain energies for a number of the [2$_n$]cyclophanes using a π-SCF force field method [152]. Lindner's predicted strain energies are presented in Table 2, and it should be noted that the agreement between calculated and experimental values is quite good.

Although many of the chemical properties of [2$_n$]cyclophanes are related to the effect of strain energy, it should be realized that it is not the total strain energy that is governing, but rather the relief of strain resulting from the breaking of a particular bond. Thus, the commercial importance of [2$_2$](1,4)cyclophane (4) is based on its

Table 2. Strain Energies of Some [2$_n$]Cyclophanes

[2$_n$]Cyclophanes	Structures	Strain Calc [152]	Energies (kcal/mol) Exp. [149-151]
anti-[2$_2$](1,3)-	(3)	15	12
[2$_2$](1,4)	(4)	28	31
[2$_2$](1,3)(1,4)-	(13)	22	23
[2$_3$](1,2,4)-	(6)	29	
[2$_3$](1,3,5)	(7)	35	
[2$_4$](1,2,4,5)	(10)	36	
[2$_6$](1,2,3,4,5,6)-	(12)	60	

thermal cleavage in the gas phase at 550 °C and 0.5 mm pressure to give a *p*-xylylene-diradicaloid intermediate (155) which, on condensing on a cool surface, gives a thin, tough, polymer coating ((157), Parylene) [19].

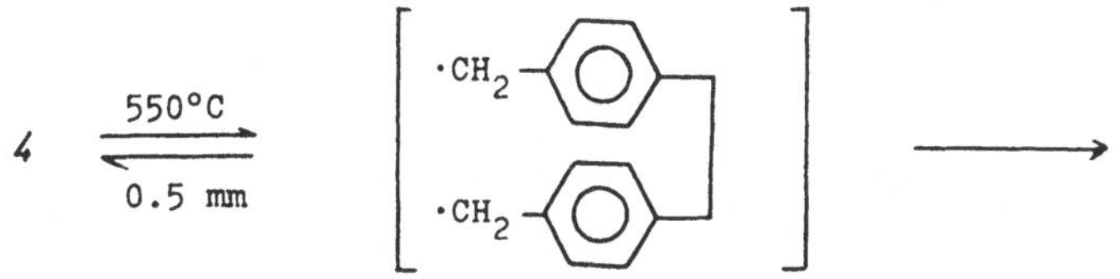

155

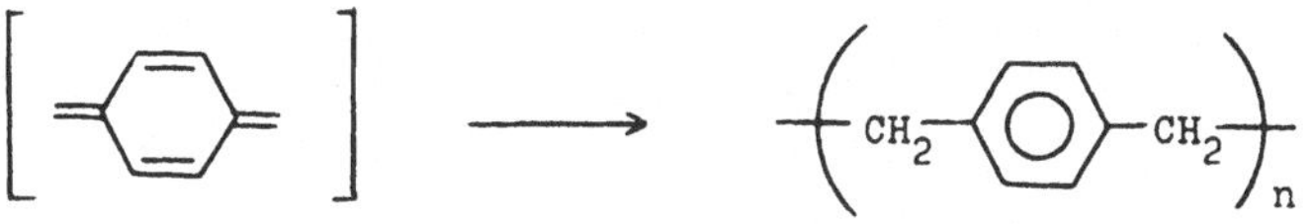

156 → 157

There is much evidence that the first step in the thermal cleavage of [2$_2$](1,4)-cyclophane (4) gives a diradical intermediate. Optically-active (—)-4-carbomethoxy-[2$_2$](1,4)cyclophane undergoes racemization following first order kinetics-(ΔG^* ~38 kcal/mol) on being heated at 200 °C in solution [153]. Similarly, thermal isomerization at 200 °C effects equilibration between pseudo-gem (158) and pseudo-meta (159), as well as between pseudo-ortho (160) and pseudo-para (161), isomers of [2$_2$](1,4)cyclophane [153].

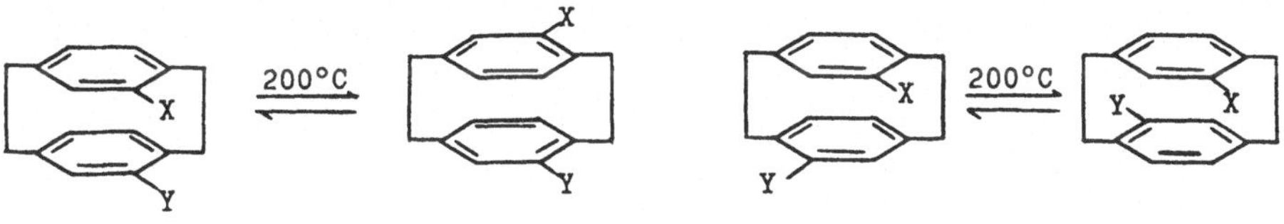

158   159   160   161

The intermediate diradical (155) can also be trapped by hydrogen donors or radical scavengers. Thus, heating (4) in the presence of hydrogen donors such as 1,4-di-(isopropyl)benzene or thiophenol leads in good yield to 1,2-bis(*p*-tolyl)ethane (162) [6,]

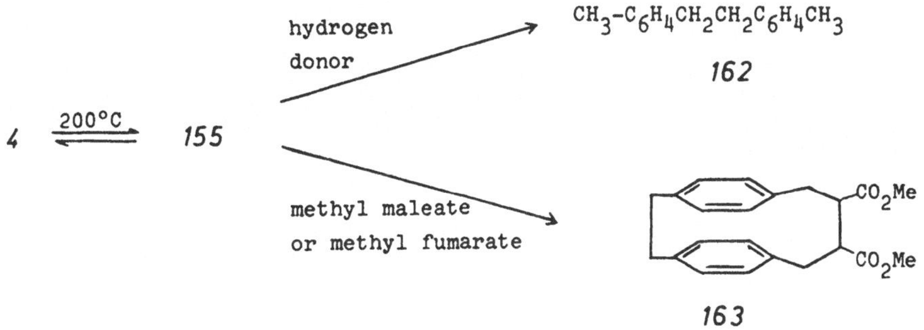

$$CH_3-C_6H_4CH_2CH_2C_6H_4CH_3$$

162

4 $\xrightarrow{200°C}$ 155

methyl maleate or methyl fumarate

163

119

[153]. The diradical intermediate (155) is also intercepted by methyl maleate or methyl fumarate, providing the ring-enlarged cyclophane (163) [153]. As would be expected for the predicated radical mechanism, the same isomeric mixture (163) is obtained from either methyl maleate or methyl fumarate.

Of the other [2₂]cyclophanes, syn-[2₂](1,3)cyclophanes (2) show a similar thermal cleavage and isomerization to give anti-[2₂](1,3)cyclophanes (3) [37]. However, the others, (1), (3), and (13) are all thermally stable. Among the higher-bridged [2ₙ]-cyclophanes, the occurrence of thermal cleavage followed by isomerization or trapping of intermediate diradicals seems dependent on whether thermal cleavage of a single bridge can lead to an unfolding of the cyclophane geometry. Thus, pyrolysis of (6) at 220 °C in thiophenol (or 1,4-di(isopropyl)benzene) gives 6,13-di-methyl[2₂](1,2)cyclophane (164) in good yield [6,68,79], and (6) likewise undergoes ring-enlargement on heating with diethyl maleate or fumarate. Similarly, heating (9) at 275 °C in 1,4-di(isopropyl)benzene gives 7,14-dimethyl[2₃](1,2,3)cyclophane (175) [79].

$$6 \quad \xrightarrow[\text{C}_6\text{H}_5\text{SH}]{220°\text{C}} \qquad 164 \qquad\qquad 9 \quad \xrightarrow{275°\text{C}} \qquad 165$$

The other [2ₙ]cyclophanes are thermally stable. Even though the multibridged cyclophanes have increased total strain, the strain energy per bridging ethano unit is less. It may also be that, if rupture of a single bridge does not allow an unfolding of the cyclophane geometry, the intermediate diradical is not accessible for trapping or polymerization. Thus, no net change occurs. For example, [2₅](1,2,3,4,5)cyclophane (11) and superphane are completely recovered unchanged after prolonged heating at 350 °C in the presence of trapping agents [81,83].

Radical cleavage of ethano bridges can also be effected photochemically. Irradiation of (4) in a glassy matrix of 2-methyltetrahydrofuran at 77 °K generates the diradical intermediate (155) and p-xylylene (156), as evidenced from their ultraviolet and fluorescence spectra [154]. Similarly, irradiation of (9) gives the diradical intermediate (166) and the p-xylylene derivative (167). The other multibridged [2ₙ]cyclophanes that have been examined, including (10), (11), and (12), are unaffected by irradiation.

$$9 \quad \xrightarrow[77°\text{K}]{h\nu} \qquad 166 \qquad \longrightarrow \qquad 167$$

Irradiation of (4) at 254 nm in alcohol gives the corresponding ring-opened ethers (168). However, surprisingly, irradiation of (4) in acetone with light of longer

wavelength than 270 nm leads to rupture of the bridge at the aromatic ring giving
(169) [155].

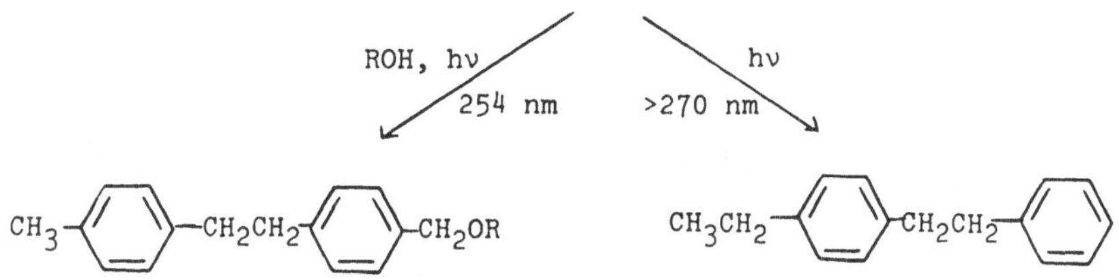

3.6.2 Electrophilic Substitution

The extensive studies of Cram on [2₂](1,4)cyclophane (4) have demonstrated that
typical electrophilic substitution reactions such as bromination, nitration, and
Friedel-Crafts acylation readily occur [57], and a similar behaviour is found for all
of the other [2ₙ]cyclophanes having unsubstituted aromatic positions. However,
in contrast to the usual electrophilic substitution of arenes, where formation of the
σ-intermediate is the slow step, loss of a proton from the σ-intermediate is generally
the slow step in electrophilic substitution of [2ₙ]cyclophanes. In part this may be
due to the steric hindrance of approach of Lewis bases in solution, but it is also
evident that the aromatic π-electron cloud of the opposite deck can serve as an
internal base. This has been established by isotope labeling experiments [57]. Thus,
bromination of (170) leads to a bromo derivative (173) in which internal transfer
between decks of the deuterium atom has occurred.

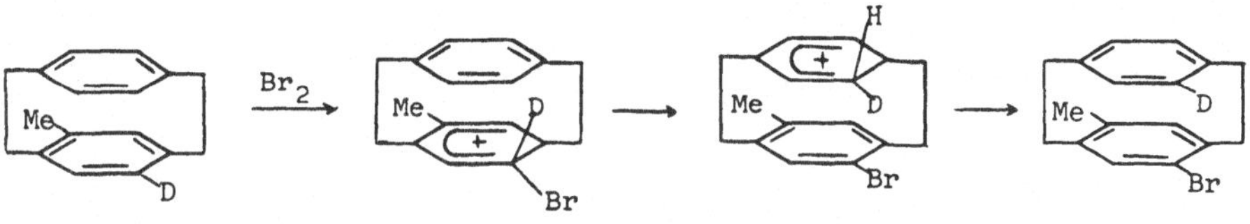

When the [2ₙ]cyclophane has a substituent that is a better base than the aromatic
π-electron cloud, substitution occurs exclusively or preferentially at the pseudo-gem
position [57], whereas in the absence of such a basic substituent the orientation of the
incoming electrophile is largely random. Thus, a carbonyl group in an ester or ketone
is a very effective base for directing electrophilic substitution pseudo-gem. This

174, R = -CH₃ or -OCH₃   175, R· = -CH₃ or -OCH₃

is shown for the conversion of (174) to (175) and provides an excellent method for making intermediates constructing additional bridges in an existing [2$_n$]cyclophane [53,63,68].

Ipso attack at bridgehead positions of [2$_n$]cyclophanes has been observed, particularly during Friedel-Crafts acylation [87,156,157]. Striking examples of ipso attack with rearrangement during acylation are the conversions of [2$_2$](1,3)(1,4)-cyclophane (13) to (176) [87] and [2$_4$](1,2,3,5)cyclophane (9) to (177) [157].

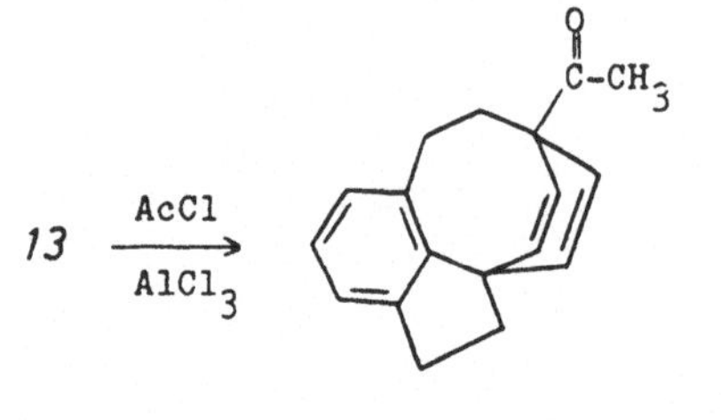

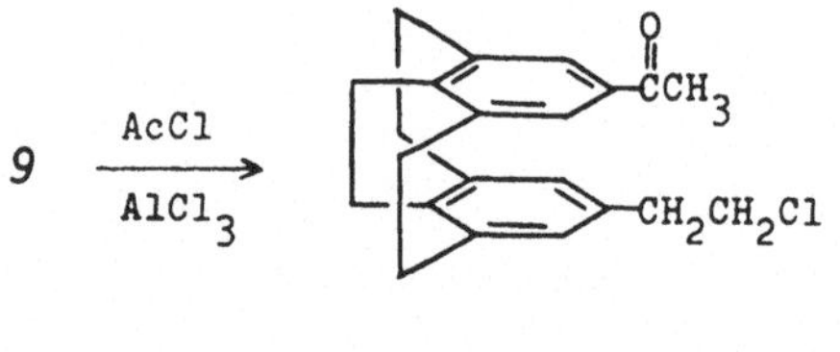

176   177

Ipso attack with rearrangement on treating cyclophanes with hydrogen chloride and aluminium chloride has been described previously as a useful method for synthesizing [2$_2$](1,3)(1,4)cyclophane (13) and [2$_3$](1,2,4)(1,2,5)cyclophane (14). Treatment of [2$_3$](1,3,5)cyclophane (7) with hydrogen chloride and aluminium chloride leads to a mixture of chloroisomers (181), whose cage structure is revealed on removal of the chlorine atoms by reduction to (182) using lithium in *tert*-butanol [51]. Presumably initial ipso attack by a proton at an aromatic bridgehead carbon is followed by transannular carbon-carbon bond formation between decks and addition of chloride ion to the opposite deck as shown by (178), (179), and (180). A double repetition of this sequence then leads to the saturated cage structures, (181) and (182) [51].

A remarkable rearrangement fostered by ipso attack is the interconversion of methyl-substituted [2$_2$](1,4)cyclophanes. Treatment of either 4,5,7,8-tetramethyl-

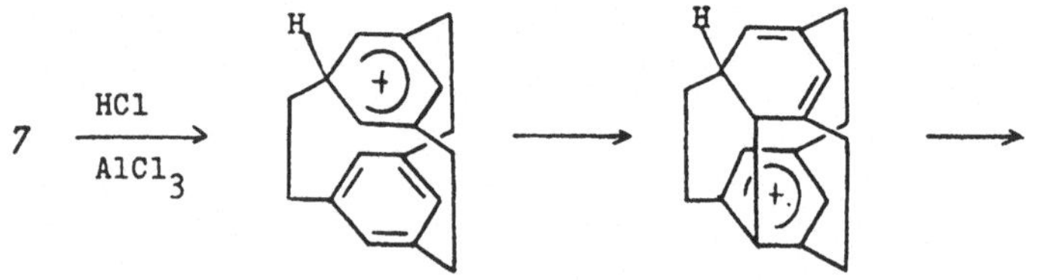

178   179

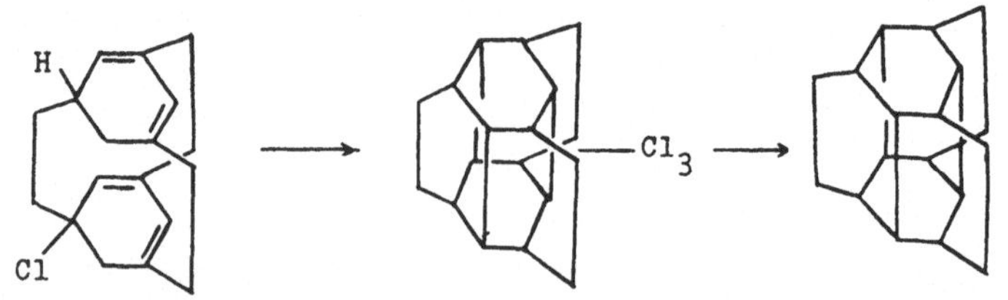

180   181   182

[2$_2$](1,4)cyclophane (183) or 4,5,12,13-tetramethyl[2$_2$](1,4)cyclophane (185) with titanium tetrachloride and hydrogen chloride at 0 °C leads in high yield to 4,7,13,16-tetramethyl[2$_2$](1,4)cyclophane (184) [158]. Deuterium labeling experiments established that methyl migration between decks is intramolecular, presumably involving a series of equilibrium steps with intermediates such as (186) and (187). This is analogous to the proton transfer between decks discussed earlier with regard to the mechanism of electrophilic substitution in [2$_n$]cyclophanes. It would be of interest to examine this acid-catalyzed alkyl transfer with a chiral alkyl group to see whether the transfer proceeds with inversion or retention of configuration. However, synthesis of an appropriate molecule for testing this question appears to be a formidable task.

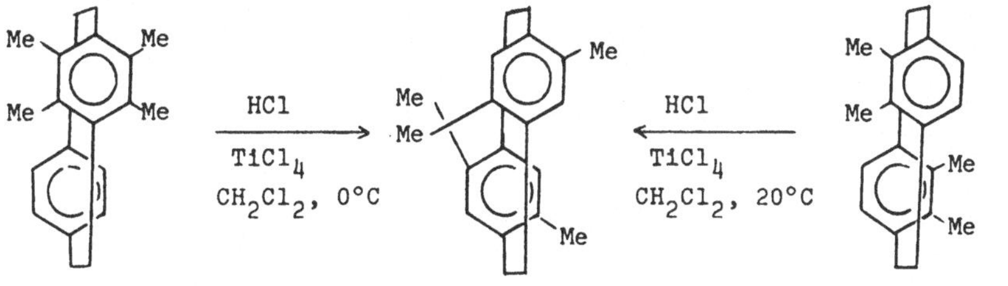

3.6.3 Reduction

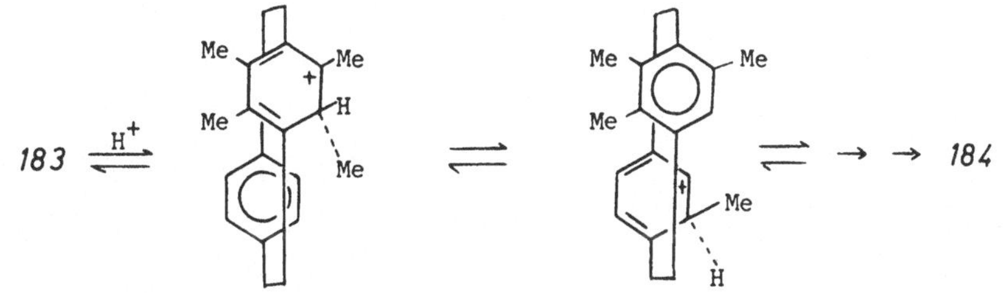

Catalytic hydrogenation of [2$_2$](1,4)cyclophane (4) in an acetic acid-ethyl acetate mixture over Adams catalyst at room temperature readily reduces the aromatic rings to give an octahydro derivative which, on prolonged hydrogenation, then yields the saturated analog (188) [159]. Examination of molecular models suggests that complete saturation of [2$_n$]cyclophane should lead to a marked increase in ring strain, particularly for the multibridged cases. Thus, it is not surprising that none of the other [2$_n$]cyclophanes have been successfully reduced to their saturated analogs. For example, [2$_3$](1,2,4)cyclophane (6) is readily hydrogenated over Adams catalyst at room temperature to give the diene (189). Prolonged hydrogenation of (189) at 70 °C led to the olefin (190), but further hydrogenation was unsuccessful [66,77].

Birch reduction of [2$_n$]cyclophanes generally relieves strain and occurs with great ease, leading to reduction of both aromatic decks. The only exception is [2$_5$](1,2,3,4,5)-cyclophane (11), which fails to react under the normal Birch conditions [81]. In the case of [2$_2$](1,4)cyclophane (4), its Birch product has the diene systems crossed, as shown by (191), rather than parallel [160–163]. Similarly, the Birch reduction of

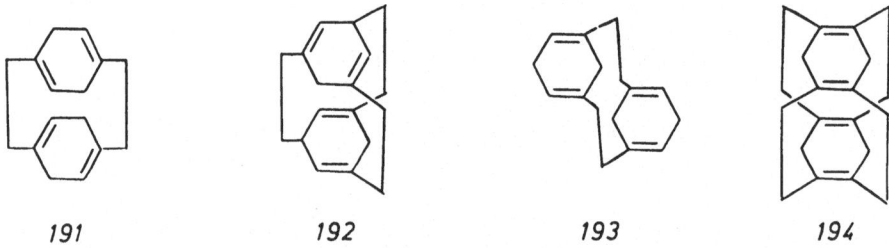

188

189   **190**

[2₃](1,3,5)cyclophane (7) yields only the crossed diene product (192) [51]. On the other hand, *anti*-[2₂](1,3)cyclophane (3) gives the Birch product (193) with parallel diene systems [164], and [2₄](1,2,4,5)cyclophane (10) is reduced in quantitative yield similarly to give a product with parallel diene systems (194) [53].

191   **192**   **193**   **194**

There is no obvious correlation between the bridging pattern of the [2ₙ]cyclophane and whether crossed or parallel diene systems will be present in the Birch product. Furthermore, [2₄](1,2,3,5)cyclophane (9) gives (195) and/or (196) [157]. In superphane (12), converting the aromatic sp² to aliphatic sp³ carbons must invoke a large increase in strain. Thus, it is not surprising that the normal Birch reduction of superphane leads mostly to recovery of unchanged superphane with only a small conversion to the mono-Birch product (197) [83].

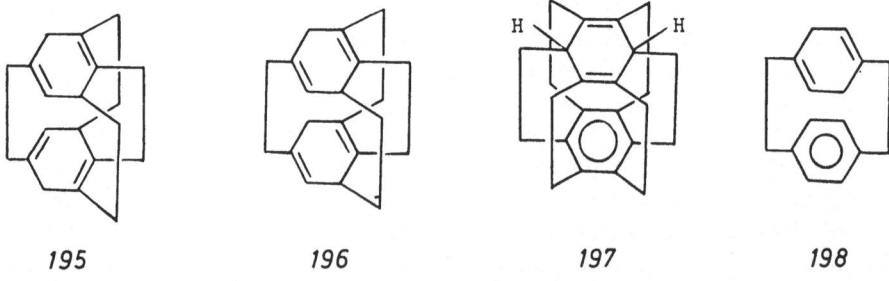

195   **196**   **197**   **198**

What is surprising is that mono-Birch products are rarely seen in these reductions, even when deliberate attempts are made to prepare the mono-Birch product by using a large excess of the [2$_n$]cyclophane. In the case of [2$_2$](1,4)cyclophane (4), a small amount of the mono-Birch product (198) is isolated [160]. However, this may be a result of thermal aromatization from the normal Birch product (191), since Jenny and Reiner have shown that (191) undergoes a ready thermal reversion to (198) [160]. It may well be that mono-Birch products are not intermediates in the formation of the normal, double-Birch products. In general the [2$_n$]cyclophanes behave as one π-electron system and it is nor obvious what the intermediates are in these Birch reductions.

However, it is a common experience that, during storage, Birch products of [2$_n$]cyclophanes will rearomatize. This occurs with especial ease, both thermally and by contact with air, in the case of superphane [83].

As shown, when more vigorous conditions (Li, EtNH$_2$, n-PrNH$_2$) were applied for the Birch reduction of superphane (12), rupture of the bridges occurs to give mainly (199) and (200), whose structures were established by independent synthesis [83].

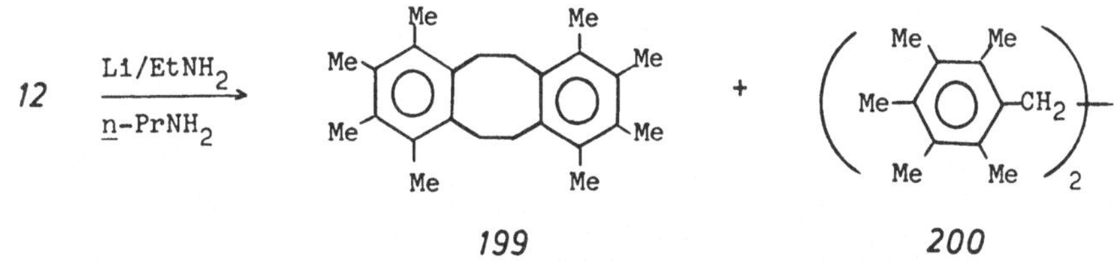

199   *200*

An interesting speculation is that electron addition to superphane yields a diradical, such as (201) which then goes on to give the bridge-ruptured products (199) and (200). An amazing reaction of superphane (12) is its conversion on treatment in concentrated sulfuric acid with an excess of zinc to the sulfur-bridged product (202) in 32% yield [83]. This reaction, which is apparently unique to superphane, may also involve a diradical intermediate such as (201) which combines with sulfur (formed from the reaction of zinc with sulfuric acid) to give (202).

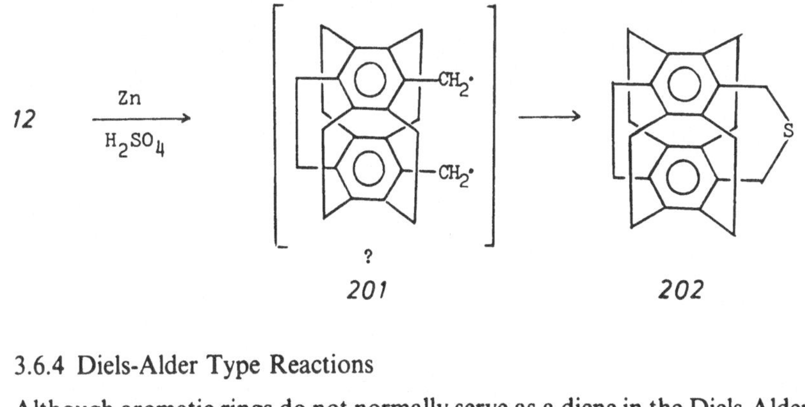

201   *202*

3.6.4 Diels-Alder Type Reactions

Although aromatic rings do not normally serve as a diene in the Diels-Alder reaction, relief of strain provides a driving force for such reactions in the case of [2$_n$]cyclophanes and the first example was that of Ciganek, who found that $[2_2](1,4)$cyclophane (4) underwent addition of dicyanoacetylene to give both the mono- and bis(barrelene) adducts (203) and (204) [165]. Since then a number of dienophiles, including tetra-cyanoethylene [53,157,166,167], dicyanoacetylene [53,81,157,167,168], hexafluoro-2-butyne [53,157,167], dimethyl acetylenedicarboxylate [67,68,166], maleic anhydride [157], and 4-phenyl-1,2,4-triazoline-3,5-dione [157], have been observed to give Diels-Alder adducts with $[2_n]$cyclophanes including (4) [165], (6) [67,68,157,166], (10) [53], and (11) [81].

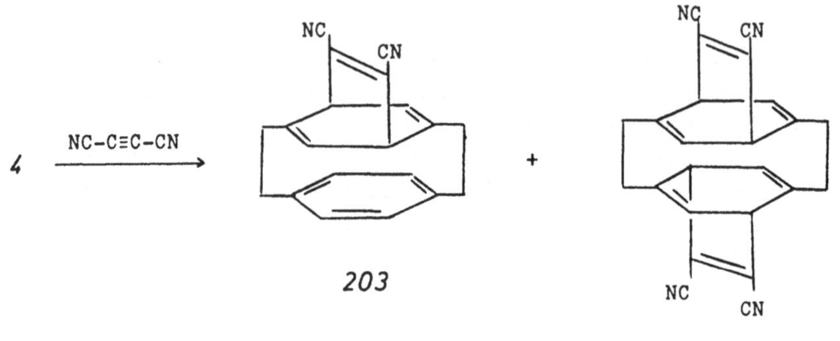

$$4 \xrightarrow{\text{NC–C≡C–CN}}$$

203 + 204

Since formation of the Diels-Alder mono-adduct is accompanied by relief of strain, addition of the second molecule of dienophile becomes more difficult and requires more strenuous conditions. Thus, by choice of conditions it is usually possible to direct the reaction to either a mono- or diadduct as desired. As before, with the more highly bridged members transformation of aromatic sp^2 to aliphatic sp^3 carbons leads to increased strain. Thus, $[2_5](1,2,3,4,5)$cyclophane (11) forms a monoadduct but not a diadduct [81]. With superphane (12), no reaction occurs even with the most reactive dienophiles under quite strenuous conditions. In an attempt to force a Diels-Alder reaction with superphane, aluminium chloride was added to catalyze the reaction with dicyanoacetylene [83]. A new product was formed in 40% yield whose structure was shown by single-crystal, x-ray analysis to be (206) [168]. Under these conditions, apparently, [2 + 2] cycloaddition occurs preferentially to give (205), which undergoes an internal Diels-Alder reaction and then adds hydrogen chloride to yield the observed product (206).

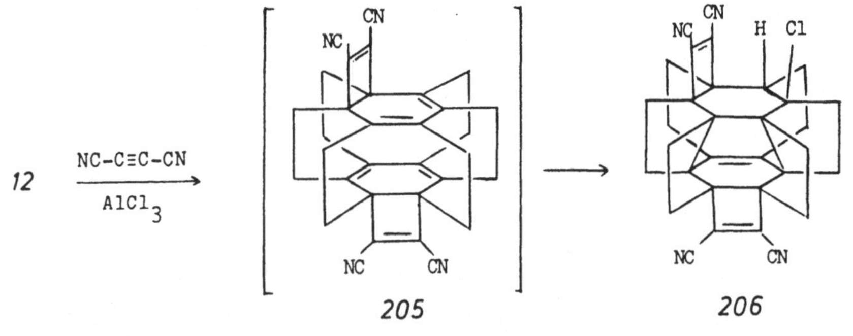

$$12 \xrightarrow[\text{AlCl}_3]{\text{NC–C≡C–CN}}$$

205   206

The aluminium chloride-catalyzed, [2 + 2] cycloaddition of dicyanoacetylene does not appear to be a general reaction of [2ₙ]cyclophanes, although it does occur with [2₂](1,4)cyclophane (4) to give the cyclooctatetraene derivative (207) [167]. As shown in Scheme 9, reduction of (207) with diisobutylaluminium hydride gives the corresponding dialdehyde (208) which, in turn, undergoes decarbonylation with tris(triphenylphosphine)rhodium(I) chloride to afford the hydrocarbon (209) [167]. Single-crystal, x-ray analyses of (207) and (209) have shown that in each case the cyclooctatetraene ring is tub-shaped rather than planar [169,170].

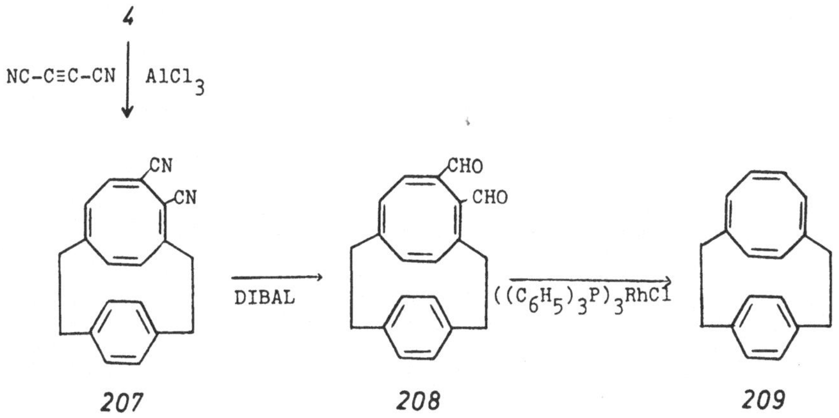

Scheme 9

Irradiation of the Diels-Alder barrelene adducts provides another method for making cyclophanes with cyclooctatetraene decks. Thus, addition of dicyanoacetylene to [2₄](1,2,4,5)cyclophane (10) gives in high yield the monoadduct (210) which, on irradiation, yields the cyclooctatetraene derivative (211) [167]. As before, (211) can be converted by reduction and decarbonylation to the corresponding hydrocarbon (212). Again single-crystal, x-ray analyses have established the structure of (213) and have shown the cyclooctatetraene ring in (212) to be tub-shaped [171,172]. In contrast to almost all other known cyclooctatetraene derivatives, (211) is unstable relative to its bicyclo[6.4.0] valence tautomer (213) and isomerizes on standing or when heated. Much of the driving force for this valence tautomerization must be the conjugative stabilization of the dicyano-diene system present in (213), because the corresponding hydrocarbon (212) shows no tendency to undergo valence tautomerization to give (214) [167].

Analogous to the Diels-Alder reactions, certain of the more reactive [2ₙ]cyclophanes undergo singlet oxygen additions. Thus, [2₄](1,2,4,5)cyclophane (10) readily adds singlet oxygen to give the epidioxide (215) [53]. From this a whole series of derivatives can be made. For example, treatment of (215) with base gives the stable hydroxydienone (216), a tautomer of the corresponding hydroquinone, and this in turn can be oxidized to the quinone (217) [53].

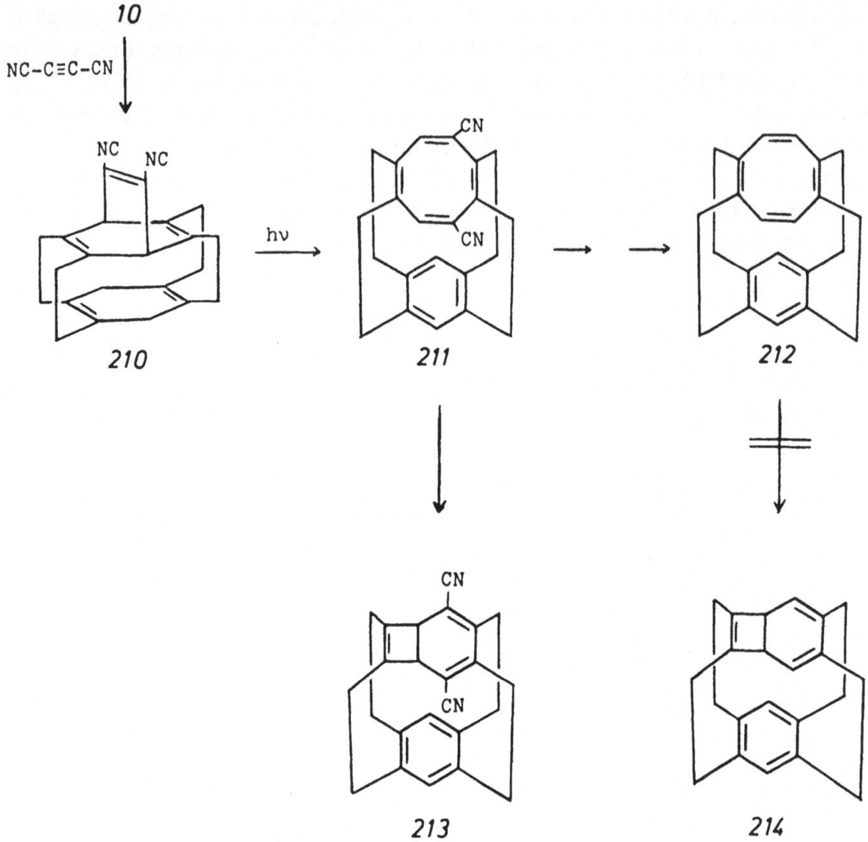

Scheme 10

Similarly, de Meijere et al. have observed singlet oxygen addition to [2₂](1,4)-cyclophane-1,9-diene (42) to give (218) [173]. Epoxidation of (218) leads mainly to (219) which, by a cobalt-catalyzed rearrangement, gives the very interesting trisoxa-tris(σ-homo)benzene derivative (220), as a thermally stable product [173].

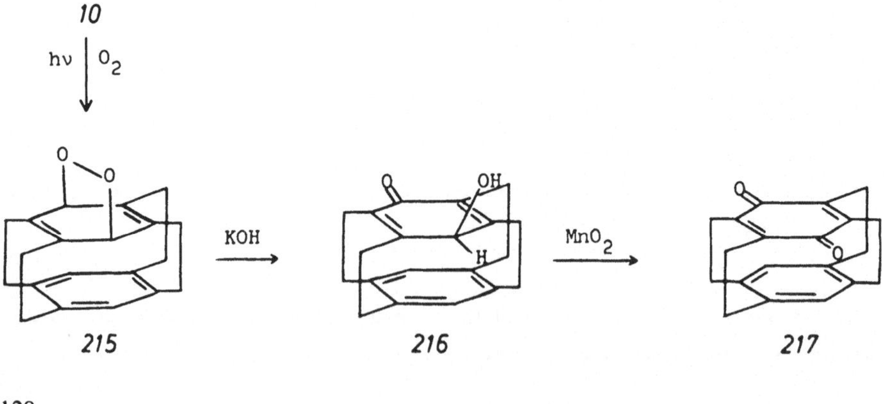

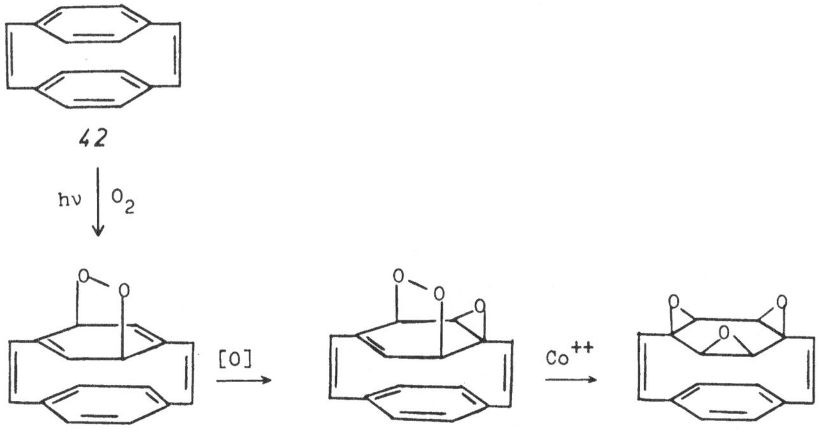

42

hν $\bigg|$ O_2

218   [O] →   **219** Co^{++} →   **220**

3.6.5 Carbene Additions

Treatment of [2₂](1,4)cyclophane (4) with diazomethane in the presence of cuprous chloride gives in poor yield a mixture of methylenated products, (221) and (222) [174,175]. With the tetra-substituted analogs, (223) and (224), the more highly methylenated products, (225) and (227), or (226) and (228), are isolated [157,176].

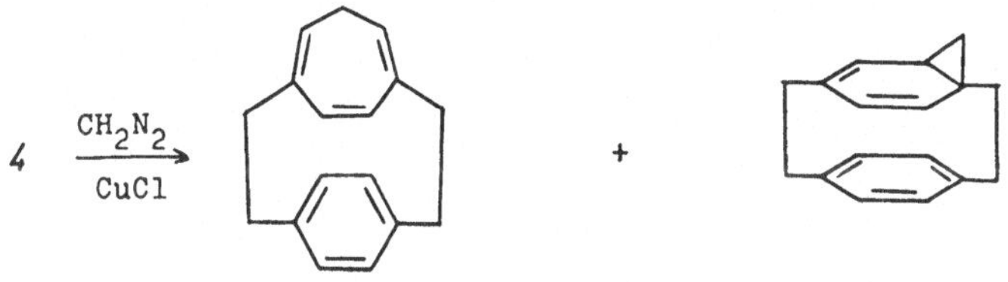

4 $\xrightarrow[\text{CuCl}]{CH_2N_2}$

221   +   **222**

$\xrightarrow[\text{CuCl}]{CH_2N_2}$ +

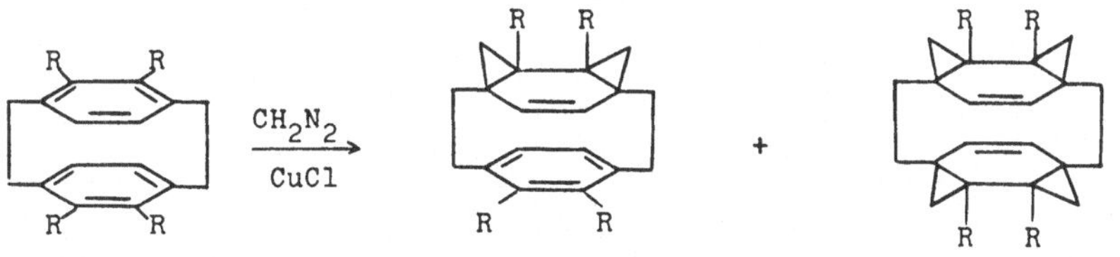

223, R = -CO₂Me   **225**, R = -CO₂Me   **227**, R = -CO₂Me

224, R = -Me   **226**, R = -Me   **228**, R = -Me

Treatment of cycloheptatriene derivatives, such as (221), with trityl fluoroborate gives cyclophanes with a tropylium ion as one deck, and these are of interest with respect to their charge-transfer interaction between decks [175,177]. Alternatively, [2ₙ]cyclophanes can be treated with ethyl diazoacetate, and this is a cleaner, better-yield approach for marking cyclophanes with a cyloheptatriene ring as one deck. By this method both (10) and (12) have been converted to their corresponding tropylium-ion charge-transfer derivatives [53,83]. The method is illustrated for the case of superphane (12) [83].

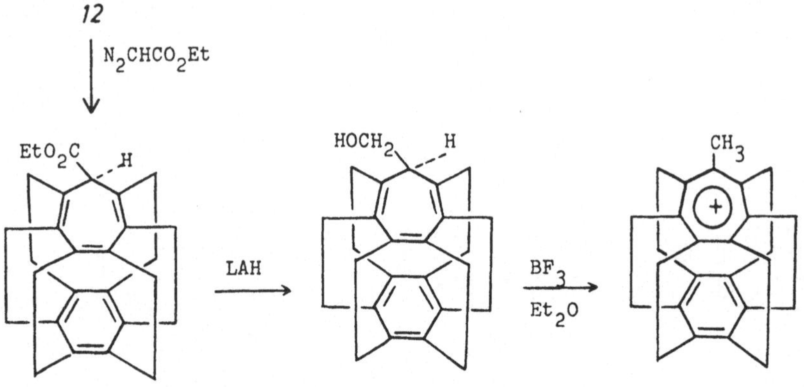

12

N_2CHCO_2Et

EtO$_2$C H

229

HOCH$_2$ H

230

LAH

CH$_3$

231

BF$_3$
Et$_2$O

As mentioned earlier, transannular carbene insertion occurs readily and is a synthetically-valuable method for constructing additional bridges in [2$_n$]cyclophanes. This reaction has figured prominently in syntheses of [2$_3$](1,2,4)cyclophane (4) [6], [2$_4$](1,2,4,5)cyclophane (10) [53], and superphane (12) [81]. A more difficult type of intramolecular carbene insertion is that reported by Boxberger et al. [178]. Pyrolysis of 232 gives (233) and (234), although in poor yield.

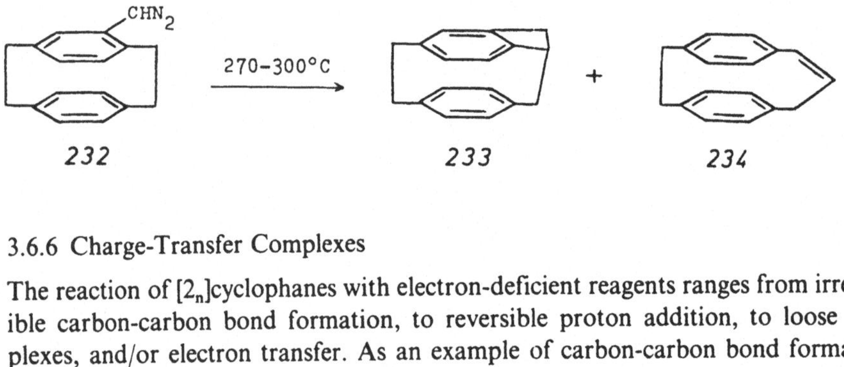

CHN$_2$

232

270–300°C

233

+

234

3.6.6 Charge-Transfer Complexes

The reaction of [2$_n$]cyclophanes with electron-deficient reagents ranges from irreversible carbon-carbon bond formation, to reversible proton addition, to loose complexes, and/or electron transfer. As an example of carbon-carbon bond formation, treatment of superphane (12) with dimethoxycarbenium fluoroborate gives the deep-red ion (235) [83]. Spectra of (235) can readily be observed in solution, and, on reduction with sodium borohydride, (235) is converted to the corresponding dihydro derivative (236). Solutions of superphane (12) in strong acid are likewise deep red, have very similar ^{1}H NMR spectra to that of (235), and on addition of base super-phane is recovered. This reversible protonation is apparently a general phenomenon of the multibridged [2$_n$]cyclophanes.

Cram and Bauer first examined the ultraviolet and visible spectra of tetracyano-ethylene (TCNE) complexes of cyclophanes, and they regarded the bathochromic shift of the long wavelength absorption band of these purple solutions to be a measure of π-electron basicity [179]. The energy level of the highest occupied molecular orbitals (π-electron basicity) in [2$_n$]cyclophanes is affected primarily by alkyl substituents, as electron donors, and by the distance between decks. With an increasing number of bridges, the distance between decks becomes shorter and so the long wavelength band of the TCNE complexes shifts to the blue. As expected, the long wavelength absorption band of superphane (12) is shifted further to the blue (572 nm) than any of the complexes of the other [2$_n$]cyclophanes [83]. However, as is clear from the analysis of the photoelectron spectra of the [2$_n$]cyclophanes [97,99], bridges do not have a simple alkyl effect, and so the shift of the long wavelength band of the TCNE complex of superphane is not nearly as large as was first anticipated.

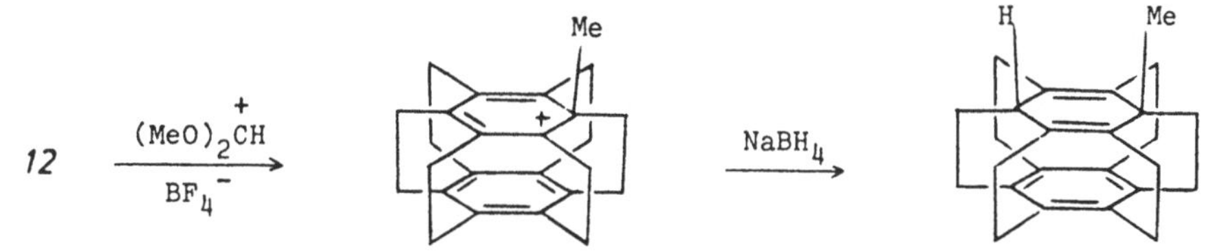

Whether the TCNE complexes are intermediates in the Diels-Alder addition reaction has not been definitely proven, but seems likely. For example, [2$_4$](1,2,4,5)-cyclophane (10) forms an immediate deep violet color with TCNE which disappears within seconds, followed by separation in quantitative yield of crystals of the corresponding Diels-Alder adduct.

The [2$_n$]cyclophane framework has been of particular value in examining between deck electron donor-acceptor interactions. Staab and Rebafka have prepared the parallel and crossed quinhydrones, (237) and (238), and have found a strong orientation effect in the charge-transfer interaction [180]. The extinction coefficient for the charge-transfer band of (237) is about ten times that of (238). Staab and his colleagues have synthesized a variety of charge-transfer complexes in the [2$_2$](1,4)-cyclophane series [181–183], and have made extensive studies of their zero field splitting parameters [184,185] and molecular orbital analyses [186]. In the [2$_2$](1,3)cyclophane series, the analogous quinhydrone (239) has been made [187], but its geometry allows only weak intermolecular hydrogen bonding rather than the strong intramolecular hydrogen bonding present in (237) and (238).

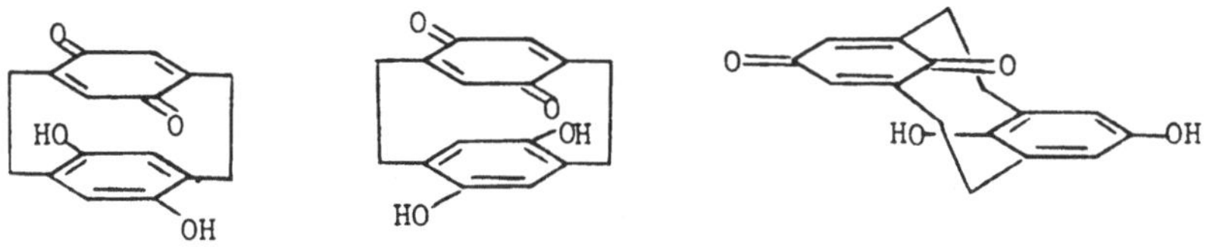

3.6.7 Transition Metal Complexes

The abundant evidence that [2$_n$]cyclophanes behave as one π-electron systems is a strong stimulus for preparing a cyclophane polymer. Such polymer molecules would also be expected to behave as a one π-electron system. Extensive synthetic work by Misumi and his colleagues has made available multilayered cyclophanes having as many as six decks as in (240) [15]. Again, the transmission effects observed for (240) and the other multilayered cyclophanes are in accord with a strong π-electron delocalization throughout the whole of the cyclophane molecule. The stepwise extension of the multilayered cyclophanes to give true polymers is an exceedingly formidable, if not impossible, task, and so new methods will need to be devised if multilayered, cyclophane polymers are to be made.

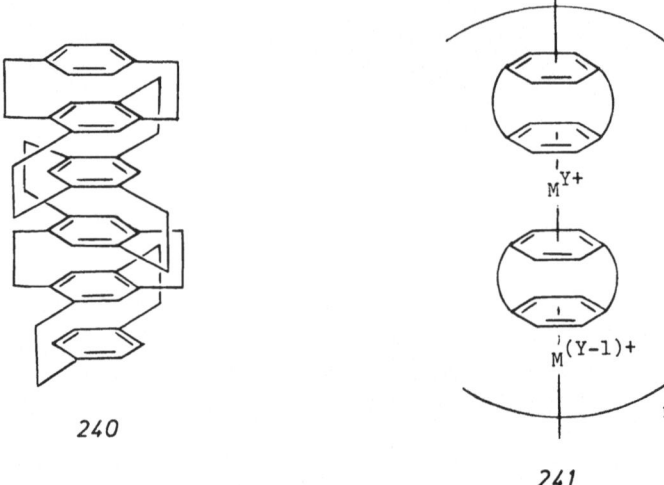

240

241

An alternate approach to cyclophane polymers is the preparation of transition metal complexes of the [2$_n$]cyclophanes which, in principle, could serve as a monomer unit for a polymer, as illustrated in a generalized structure[1] by (241). It is this concept [188] that has stimulated much of the recent surge of interest in transition metal complexes of cyclophanes.

The early work of Cram and Wilkinson demonstrated that [2$_2$](1,4)cyclophane readily reacts with hexacarbonylchromium to give (242) [189]. Since then, a number of tricarbonylchromium complexes of [2$_2$](1,4)cyclophane have been made in this way [190–192]. Misumi and co-workers have also prepared the double complex (243) [193]. By means of the metal atom technique, Elschenbroich et al. made the unusual com-

---

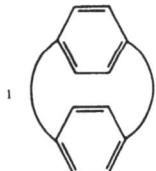

[1]   is a generalized structure intended to represent all of the possible [2$_n$]cyclophanes.

plex (244) having an internal chromium atom, and also the complex of a chromium atom with two cyclophane molecules, (245) [194].

Cr(CO)3
Cr(CO)3
Cr(CO)3
Cr
Cr+

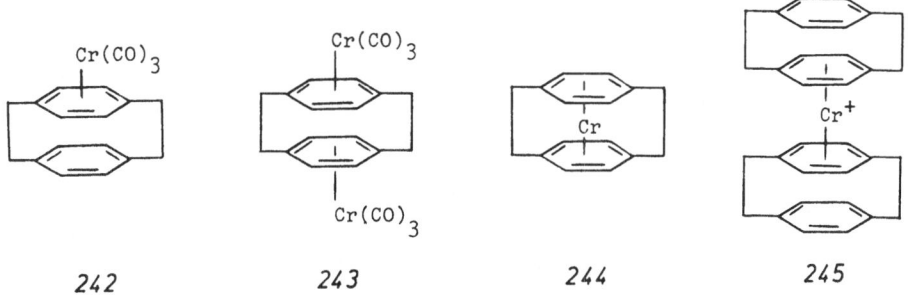

242 *243* *244* *245*

Neither the tricarbonylchromium nor the metal atom approach are suited to preparing oligomers and polymers of [2_n]cyclophane-transition metal complexes, and so attention has turned to other techniques and other metals. The first to be

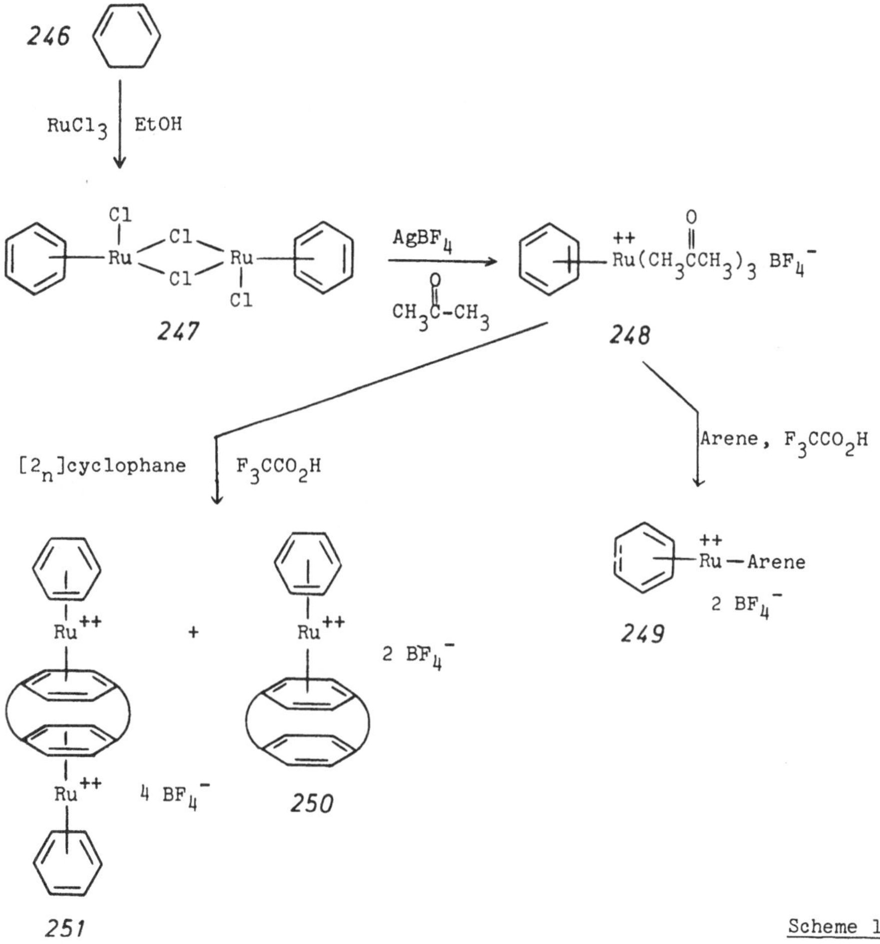

246
RuCl3
EtOH
Cl
Cl
Ru
Ru
Cl
Cl
247
AgBF4
CH3C-CH3
++
Ru(CH3CCH3)3 BF4-
248
Arene, F3CCO2H
[2n]cyclophane
F3CCO2H
++
Ru—Arene
2 BF4-
249
Ru++
+
Ru++
2 BF4-
Ru++
4 BF4-
250
251

Scheme 11

employed was the method of Bennett and his colleagues, who developed an efficient method for preparing bis(η^6-arene)ruthenium(II) complexes [195,196]. In this method, a dihydroarene, such as (246), is heated with ruthenium trichloride to form a bis(arene)dichloro-μ-chlorodiruthenium(II), (247), and this on treatment with silver tetrafluoroborate in acetone gives the solvated η^6-areneruthenium(II) complex (248). Treatment of (248) with a second arene in the presence of trifluoracetic acid leads in excellent yield to thermal- and air-stable bis(η^6-arene)ruthenium complexes (249) [195,196]. When [2_n]cyclophanes are employed as the second arene, the double- and triple-decker ruthenium complexes (250) and (251) are readily formed [197,198].

When the ruthenium solvate (248) and the [2_n]cyclophane are employed in a 1:1 ratio, the double-decker complexes (250) are obtained in excellent yield; whereas, when the ruthenium solvate (248) is used in excess, the triple-decker complexes (251) result. About twenty-five cyclophane-ruthenium complexes of the general structures (250) and (251) have been Uade, where the arene has been varied among benzene, p-cymene, mesitylene, and hexamethylbenzene and the cyclophane component has included [2_2](1,4)cyclophane (4), anti-[2_2](1,3)cyclophane (3), [2_3](1,2,4)-cyclophane (6), [2_3](1,3,5)cyclophane (7), [2_4](1,2,3,4)cyclophane (8), [2_4](1,2,3,5)-cyclophane (9), [2_4](1,2,4,5)cyclbphane (10), [2_5](1,2,3,4,5)cyclophane (11), and super-phane (12) [197,198,199].

Recently, Gill and Mann found that irradiation of (η^5-cyclopentadienyl)(η^6-benzene)ruthenium(II) cation (252) in acetonitrile gives the stable cyclopentadienyltris-(acetonitrile)ruthenium(II) cation (253) which readily combines thermally with arenes [200]. When cyclophanes are employed as the arenes, this method provides a convenient, high-yield procedure for preparing the double- and triple-decker ruthenium complexes (254) and (255) in which the capping substituent is the cyclopentadienyl anion rather than an arene [201].

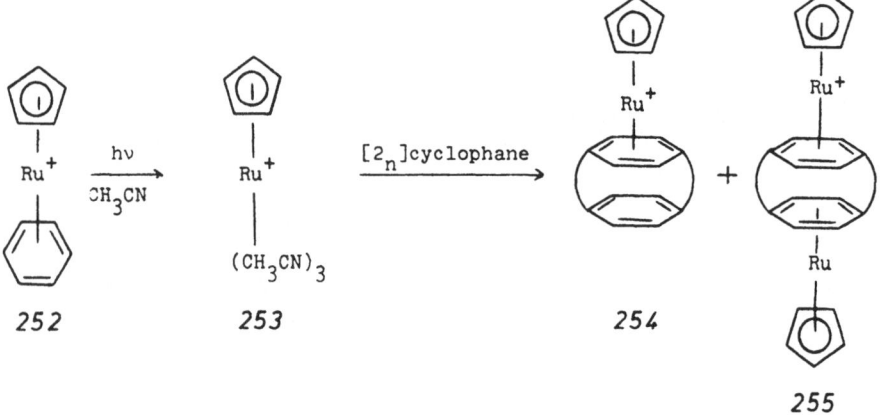

Similarly, Gill and Mann have shown that (η^5-cyclopentadienyl)(η^6-p-xylene)-iron(II) (256) readily undergoes arene exchange reactions on irradiation [202]. When

[2$_n$]cyclophanes are employed as the second arene, the double- and triple-decker iron complexes (257) and (258) are readily formed in good yield [188,199,203]. Also, the classical ferrocene-arene exchange reaction catalyzed by aluminium trichloride and aluminum [204,205] has been used for making double- and triple-decker complexes of [2$_n$]cyclophanes [188,206,207]. The iron complexes of [2$_n$]cyclophanes that have been made include *anti*-[2$_2$](1,3)cyclophane (3), [2$_2$](1,4)cyclophane (4), [2$_4$](1,2,4,5)-cyclophane (10), [2$_5$](1,2,3,4,5)cyclophane (11), and superphane (12) [199].

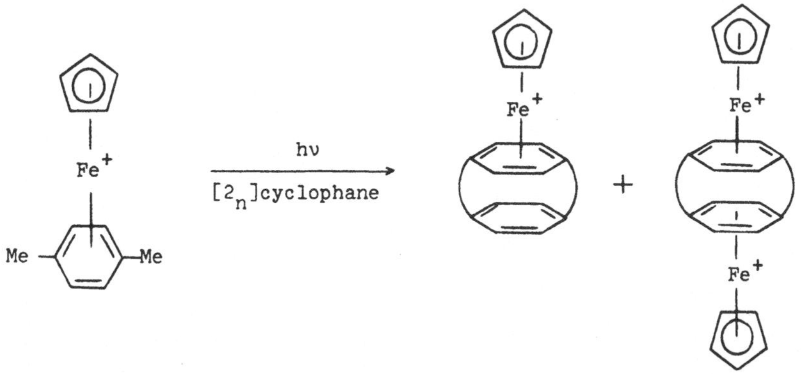

256   257   258

The iron(II) and ruthenium(II) complexes of the [2$_n$]cyclophanes are thermal- and air-stable solids which can be stored indefinitely. Their ^{1}H- and ^{13}C NMR spectra provide a useful insight into the fundamental nature of the metal-arene bonding. In particular, complexes of *anti*-[2$_2$](1,3)cyclophane (3) allow a dissection of the ring current and electron-withdrawing effects of complexation by the metal. This has been studied for the iron [208], ruthenium [198], and tricarbonylchromium [209] complexes of (3).

Cyclic voltammetry of the (η^6-arene)(η^6-[2$_n$]cyclophane)ruthenium(II) complexes (254) show in general a reversible, two-electron reduction [198]. Bulk electrolysis is a good preparative method for making the corresponding ruthenium(0) complexes (260) [210]. The bis(hexamethylbenzene)ruthenium(0) complex (259) is a flux-ional molecule having the η^6-η^4 structure shown [211,212]. It is of particular interest that, in the two examples of [2$_n$]cyclophane-ruthenium(0) complexes (where the [2$_n$]cyclophane is (4) or (10)), whose spectra have been studied in some detail, the hexamethylbenzene ring is bound η^6 [210]. This is in accord with the fact that the [2$_n$]-cyclophanes having a geometry best suited for η^4-complexation are also the ones most readily reduced.

In contrast, cyclic voltammetry of the iron(II) complexes (257) shows a reversible one-electron reduction [213]. These iron 19-electron structures (261) are an interesting addition to the category of electron-reservoir compounds, much studied by Astruc and his colleagues [214].

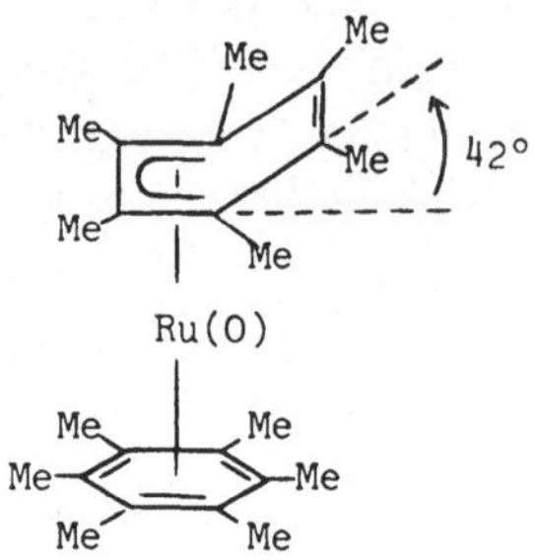

259 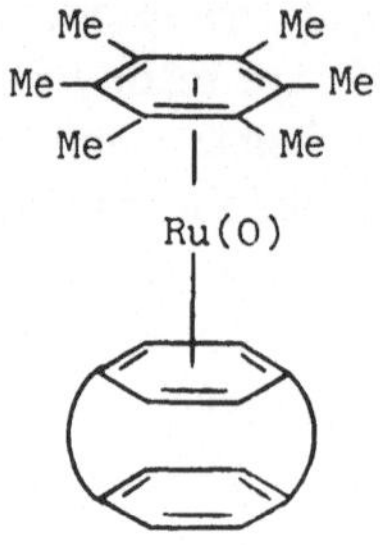   260 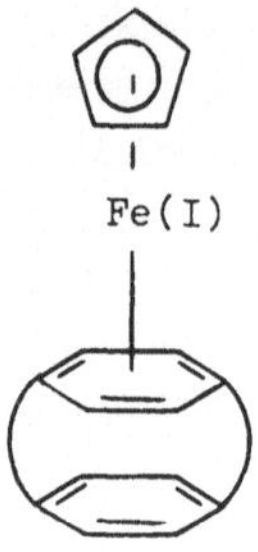   261

4 Summary

In his first paper on cyclophanes, Professor Cram made some very penetrating observations about the value of these molecules for testing bonding, ring strain, and π-electron interactions — a remarkably accurate forecast of future developments in the field. But probably, even he did not forsee the extent to which the field would develop, particularly the new synthetic methods and synthetic accomplishments which have matched the growth in our theoretical understanding.

With syntheses of all of the [2n]cyclophanes in hand, what are the likely new directions? Undoubtedly, there will continue to be much synthetic activity. One can always conceive of additional molecules that would be valuable for testing theoretical concepts. Another area will be discovering even better syntheses for making all of the [2n]cyclophanes in quantity. Presently, studies of the chemical and physical properties of the [2n]cyclophanes is limited by their availability. Only [2₂](1,4)cyclophane (4) is available commercially in unlimited research quantities.

Just as the [2n]cyclophanes have served so well for exploring and testing theoretical concepts in organic chemistry, it is to be expected that they will serve a similar role in the exploding area of organometallic chemistry. This is an area of cyclophane chemistry whose surface has hardly been scratched.

Finally, there is much consolidation that needs to be done for the already known, and unusual, [2n]cyclophane syntheses and [2n]cyclophane chemistry. In-depth, mechanistic studies are needed.

5 Acknowledgements

We thank the National Science Foundation for their support of that part of our research included in this review. The author thanks the Alexander von Humboldt-Stiftung for a reinvitation award during the tenure of which this review was written.

The author also thanks Professor Klaus Hafner and his colleagues at the Technische Hochschule Darmstadt for their gracious hospitality during this tenure period.

6 Notes Added in Proof (Dec. 6, 1982)

Klieser and Vögtle, by use of the cesium carbonate procedure [215], have successfully effected the coupling of tetrathiol (262) with tetrabromide (263) to give a mixture of the tetrathiacyclophanes (264) and (265) in a single step in 10% overall yield [216]. Conversion of the mixture of (264) and (265) to the corresponding tetrasulfones (266) and (267) followed by pyrolysis provides a three-step synthesis of the important [2$_n$]cyclophane (10), again accompanied by (79) [217].

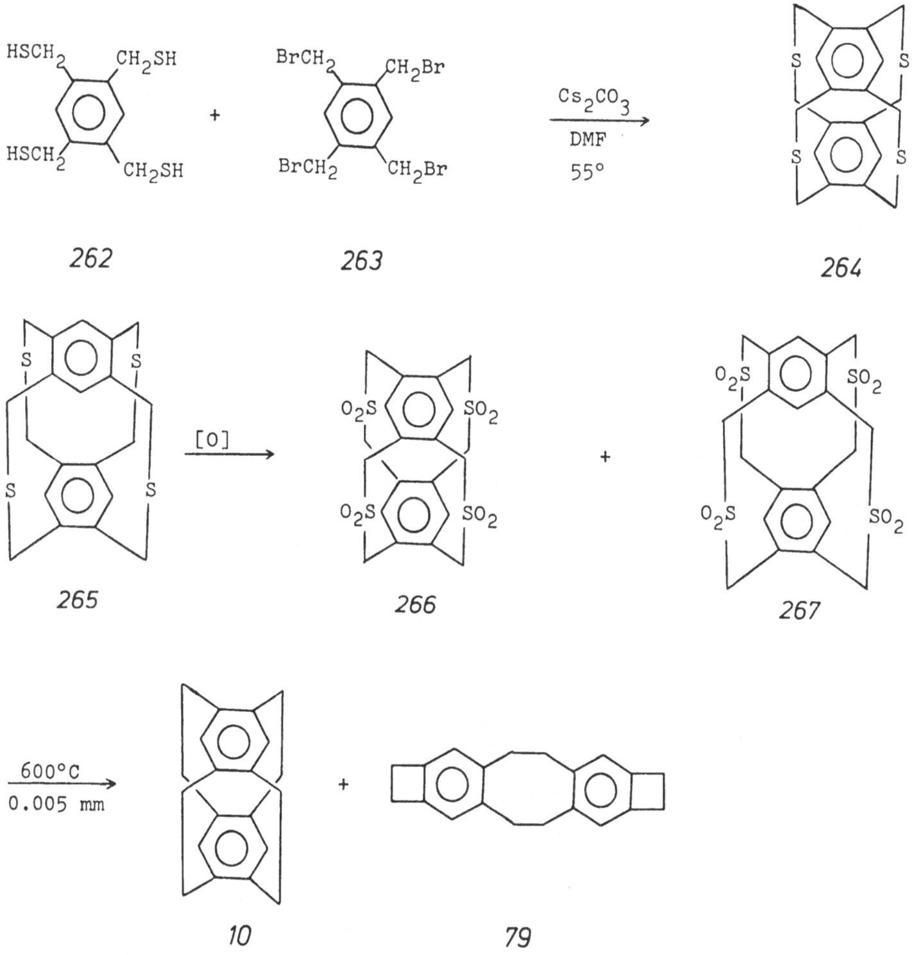

262   263   264

265   266   267

10   79

El-tamany and Hopf have described a second synthesis of superphane (12), as outlined in *Scheme 12*, starting with 4,5,12,13-tetramethyl[2$_2$](1,4)cyclophane (63) and introducing the remaining bridges stepwise as shown [218].

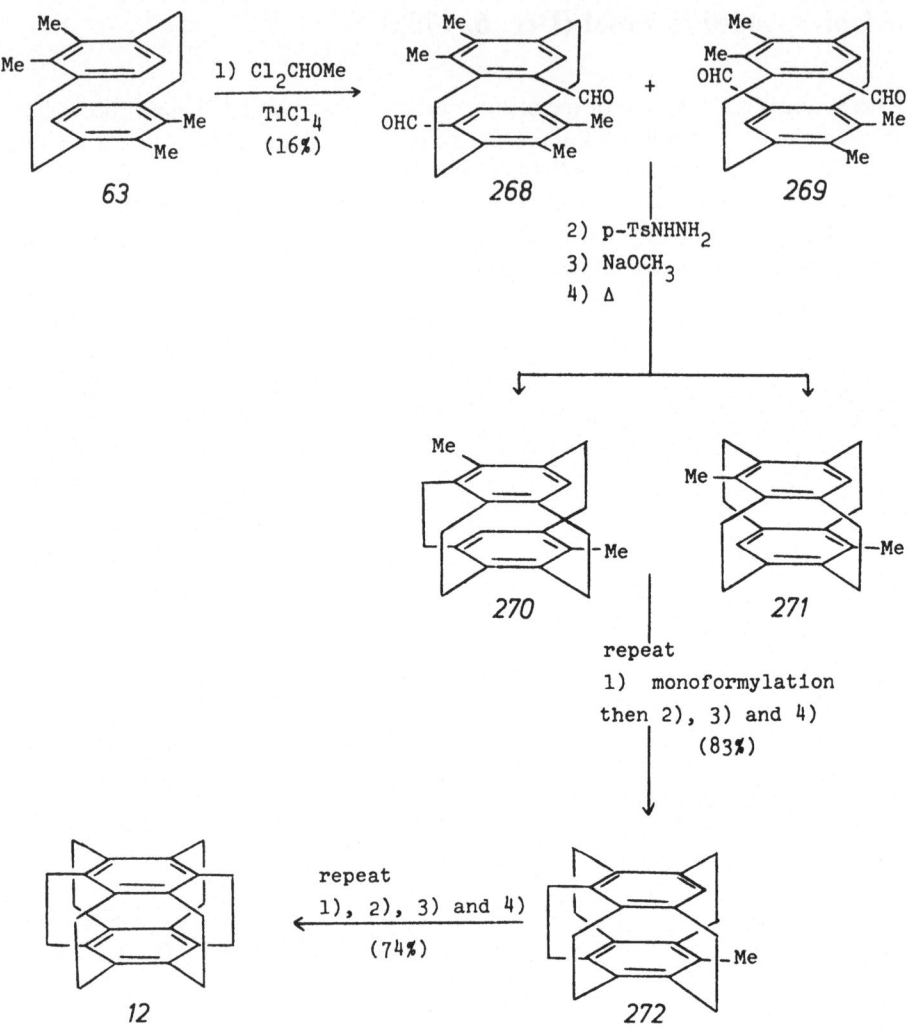

Scheme 12

7 References

1. Brown, C. G., Farthing, A. C.: Nature *164*, 915 (1949)
2. Farthing, A. C.: J. Chem. Soc. 3621 (1953)
3. Brown, C. J.: ibid. 3265 (1953)
4. Cram, D. J.: Rec. Chem. Prog. *20*, 71 (1959)
5. Cram, D. J., Cram, J. M.: Acc. Chem. Res. *4*, 204 (1971)
6. Cram, D. J., Hornby, R. B., Truesdale, E. A., Reich, H. J., Delton, M. H., Cram, J. M.: Tetrahedron *30*, 1757 (1974)
7. Cram, D. J., Steinberg, H. J.: J. Am. Chem. Soc. *73*, 5691 (1951)
8. Schubert, W. M., Sweeney, W. A., Latourette, H. K.: ibid. *76*, 5462 (1954)
9. Cram, D. J., Abell, J.: ibid. *77*, 1179 (1955)
10. Smith, B. H.: Bridged Aromatic Compounds, Academic Press, New York, 1964, pp. 1–24
11. Vögtle, F., Neumann, P.: Tetrahedron Letters *1969*, 5329; Tetrahedron *26*, 5847 (1970)
12. Lehner, H.: Monatsh. Chem. *107*, 565 (1976)
13. Misumi, S.: Mem. Inst. Sci. Ind. Res., Osaka Univ. *33*, 53 (1976)
14. (a) Vögtle, F., Neumann, P.: Top. Curr. Chem. *48*, 67 (1974);
    (b) Vögtle, F., Neumann, P.: Synthesis *1973*, 85;
    (c) Vögtle, F., Hohner, G.: Top. Curr. Chem. *74*, 1 (1978)
15. Misumi, S., Otsubo, T.: Acc. Chem. Res. *11*, 251 (1978)
16. Boekelheide, V.: ibid. *13*, 65 (1980)
17. Hopf, H.: Nachr. Chem. Tech. Lab. *28*, 311 (1980)
18. (a) Pellegrin, M.: Rec. Trav. Chim., Pays-Bas, *18*, 458 (1899);
    (b) Baker, W., Banks, R., Lyon, D. R., Mann, F. G.: J. Chem. Soc. 27 (1945)
19. (a) Gorham, W. F.: German Pat. 1,085,763, July 21, 1960; Chem. Abstr. *55*, 22920h (1961);
    (b) Gorham, W. F.: J. Polym. Sci., Part A-1, *4*, 3027 (1966)
20. (a) Winberg, H. E., Fawcett, F. S., Mochel, W. E., Theobald, C. W.: J. Am. Chem. Soc. *82*, 1428 (1960);
    (b) Winberg, H. E., Fawcett, F. S.: Organic Syntheses, Collect. Vol. *5*, Wiley, New York, 1973, p. 883
21. Otsubo, T., Horita, H., Misumi, S.: Synth. Comm. *6*, 591 (1976)
22. Ito, Y., Miyata, S., Nakatsuka, M., Saegusa, T.: J. Org. Chem. *46*, 1043 (1981)
23. Sato, T., Wakabayashi, M., Kainosho, M.: Tetrahedron Letters 4185 (1968)
24. Vögtle, F.: Angew. Chem. *81*, 258 (1969); ibid, Angew. Chem. Int. Ed. Eng. *8*, 274 (1969)
25. Vögtle, F., Schunder, L.: Chem. Ber. *102*, 2677 (1969)
26. Vögtle, F.: ibid. *102*, 3077 (1969)
27. Vögtle, F.: Tetrahedron *25*, 3231 (1969)
28. Vögtle, F.: Chemiker Zeitung *94*, 313 (1970)
29. Haenel, M., Staab, H. A.: Tetrahedron Letters 3585 (1970)
30. Mitchell, R. H., Boekelheide, V.: J. Am. Chem. Soc. *92*, 3510 (1970)
31. Mitchell, R. H., Boekelheide, V.: J. Chem. Soc., Chem. Commun. *1970*, 1555
32. Mitchell, R. H., Boekelheide, V.: Tetrahedron Letters 1197 (1970)
33. Mitchell, R. H., Boekelheide, V.: J. Am. Chem. Soc. *96*, 1547 (1974)
34. Bruhin, J., Jenny, W.: Tetrahedron Letters 1215 (1973)
35. Vögtle, F.: Chemistry and Industry *1972*, 346
36. Boekelheide, V., Tsai, C. H.: J. Org. Chem. *38*, 3931 (1973)
37. Kamp, D., Boekelheide, V.: ibid. *43*, 3470 (1978)
38. Anker, W., Bushnell, G. W., Mitchell, R. H.: Can. J. Chem. *57*, 3080 (1979)
39. Mitchell, R. H., Otsubo, T., Boekelheide, V.: Tetrahedron Letters 219 (1975)
40. Otsubo, T., Gray, R., Boekelheide, V.: J. Am. Chem. Soc. *100*, 2449 (1978)
41. Mitchell, R. H., Yan, J. S. H., Dingle, T. W.: ibid. *104*, 2551 (1982)
42. Otsubo, T., Boekelheide, V.: Tetrahedron Letters 3881 (1975)
43. Otsubo, T., Boekelheide, V.: J. Org. Chem. *42*, 1085 (1977)
44. Umemoto, T., Otsubo, T., Misumi, S.: Tetrahedron Letters 1573 (1974)
45. Boekelheide, V., Reingold, I. D., Tuttle, M.: J. Chem. Soc., Chem. Commun. *1973*, 406
46. Staab, H. A., Haenel, M.: Chem. Ber. *106*, 2190 (1973)
47. Rebafka, W., Staab, H. A.: Angew. Chem. *85*, 831 (1973); *ibid*: Angew. Chem., Int. Ed. Engl. *12*, 776 (1973)

48. Gray, R., Harruff, L. G., Krymowski, J., Petersen, J., Boekelheide, V.: J. Am. Chem. Soc. *100*, 2892 (1978)
49. Givens, R. S., Wylie, P. L.: Tetrahedron Letters 865 (1978)
50. Hibert, M., Solladie, G.: J. Org. Chem. *45*, 4496 (1980)
51. Boekelheide, V., Hollins, R. A.: J. Am. Chem. Soc. *95*, 3201 (1973); ibid. *92*, 3512 (1970)
52. Gray, R., Boekelheide, V.: Angew. Chem. *87*, 138 (1975); ibid. Angew. Chem. Int. Ed. Eng. *14*, 107 (1975)
53. Gray, R., Boekelheide, V.: J. Am. Chem. Soc. *101*, 2128 (1979)
54. Sherrod, S. A., da Costa, R. L., Barnes, R. A., Boekelheide, V.: ibid. *96*, 1565 (1974)
55. Nakazaki, M., Yamamoto, Y., Muira, Y.: J. Org. Chem. *43*, 1041 (1978)
56. Staab, H. A., Schwendemann V. M.: Liebig's Ann. Chem. *1979*, 1258
57. Reich, H. J., Cram, D. J.: J. Am. Chem. Soc. *91*, 3505, 3517, 3527 (1969)
58. Hopf, H.: Angew. Chem. *84*, 471 (1972); ibid. Angew. Chem., Int. Ed. Eng. *11*, 419 (1972)
59. Hopf, H., Lenich, F. Th.: Chem. Ber. *107*, 1891 (1974)
60. Böhm, I., Herrmann, H., Menke, K., Hopf, H.: ibid. *111*, 523 (1978)
61. Chapman, O., Ed.: Organic Syntheses, Vol. 60, J. Wiley and Sons, New York, 1981, pp. 41–48
62. Trampe, S., Menke, K., Hopf, H.: Chem. Ber. *110*, 371 (1977)
63. Gilb, W., Menke, K., Hopf, H.: Angew. Chem. *89*, 177 (1977); ibid. Angew. Chem., Int. Ed. Eng. *16*, 191 (1977)
64. Kleinschroth, J., Hopf, H.: Angew. Chem. *91*, 336 (1979); ibid. Angew. Chem., Int. Ed. Eng. *18*, 329 (1979)
65. Kleinschroth, J.: Doctoral Dissertation, Wurzburg, 1980
66. Murad, A. E., Hopf, H.: Chem. Ber. *113*, 2358 (1980)
67. Truesdale, E. A., Cram, D. J.: J. Am. Chem. Soc. *95*, 5825 (1973)
68. Truesdale, E. A., Cram, D. J.: J. Org. Chem. *45*, 3974 (1980)
69. Jacobson, N., Boekelheide, V.: Angew. Chem. *90*, 49 (1978); ibid. Angew. Chem., Int. Ed. Eng. *17*, 46 (1978)
70. Eltamany, S. H., Hopf, H.: Tetrahedron Letters *21*, 4901 (1980)
71. Longone, D. T., Simanyi, L. H.: J. Org. Chem. *29*, 3245 (1964)
72. Lindberg, T., Ed.: Strategies and Tactics of Organic Synthesis, Chapter XX, V. Boekelheide, J. Wiley and Sons, New York, 1983
73. Boekelheide, V., Ewing, G.: Tetrahedron Letters *1978*, 4245
74. Simultaneously with our own work, Prof. Schiess developed the pyrolysis of o-chloromethyltoluenes as a general method for making benzocyclobutenes; see, Schiess, P., Heitzmann, M.: Helv. Chim. Acta *61*, 844 (1978); and Schiess, P., Heitzmann, M., Rutschmann, S., Stäheli, R.: Tetrahedron Letters 4569 (1978)
75. (a) Cava, M. P., Deana, A. A.: J. Am. Chem. Soc. *81*, 4266 (1959);
    (b) Laganis, E. D., Boekelheide, V.: Unpublished Work
76. Ewing, G. D., Boekelheide, V.: J. Chem. Soc., Chem. Commun. *1979*, 207
77. Aalbersberg, W. G. L., Vollhardt, K. P. C.: Tetrahedron Letters *1979*, 1939
78. Neuschwander, B., Boekelheide, V.: Israel J. Chem. *20*, 288 (1980)
79. Hopf, H., Kleinschroth, J., Murad, A. E.: ibid. *20*, 291 (1980)
80. Schirch, P. F. T., Boekelheide, V.: J. Am. Chem. Soc. *101*, 3125 (1979)
81. Schirch, P. F. T., Boekelheide, V.: ibid. *103*, 6873 (1981)
82. Sekine, Y., Boekelheide, V.: ibid. *101*, 3126 (1979)
83. Sekine, Y., Boekelheide, V.: ibid. *103*, 1777 (1981)
84. Rieche, A., Gross, H., Höft, E.: Chem. Ber. *93*, 88 (1960)
85. For similar transannular carbenium ion-hydride abstractions to form pseudo-gem methyl aldehydes, see references 6 and 53
86. Cram, D. L., Helgeson, R. C., Lock, D., Singer, L. A.: J. Am. Chem. Soc. *88*, 1324 (1966)
87. Hefelfinger, D. T., Cram, D. J. ibid. *93*, 4754, 4767 (1971)
88. Hopf, H.: Private Communication
89. Gilman, R. E., Delton, M. H., Cram, D. J.: J. Am. Chem. Soc. *94*, 2478 (1972)
90. Boekelheide, V., Anderson, P. H., Hylton, T. A.: ibid. *96*, 1558 (1974)
91. Boekelheide, V., Tsai, C. H.: J. Org. Chem. *38*, 3931 (1973)

92. Juriew, J., Skorochodowa, T., Merkuschew, J., Winter, W., Meir, H.: Angew. Chem. *93*, 285 (1981); ibid. Angew. Chem., Int. Ed. Engl. *20*, 269 (1981)
93. Seebach, D., Jones, N. R., Corey, E. J.: J. Org. Chem. *33*, 300 (1968)
94. Hylton, T., Boekelheide, V.: J. Am. Chem. Soc. *90*, 6887 (1968)
95. Gschwend, H. W.: ibid. *94*, 8430 (1972)
96. Bickert, P., Boekelheide, V., Hafner, K.: Angew. Chem. *94*, 308 (1982); ibid. Angew. Chem., Int. Ed. Engl. *21*, 304 (1982)
97. Kovac, B., Mohraz, M., Heilbronner, E., Boekelheide, V., Hopf, H.: J. Am. Chem. Soc. *102*, 4314 (1980)
98. Cram, D. J., Wechter, W. J., Kierstad, R. W.: ibid. *80*, 3126 (1958)
99. Heilbronner, E.: Top. Curr. Chem. *115*, 1 (1983)
100. Gerson, F.: Top. Curr. Chem. *115*, 57 (1983)
101. Brown, C. J.: J. Chem. Soc. *1953*, 3278
102. Kai, Y., Yasuoka, N., Kasai, N.: Acta Crystallogr. *B33*, 754 (1977)
103. Hanson, A. W.: ibid. *15*, 956 (1962)
104. Mathew, M.: ibid. *B24*, 530 (1968)
105. Hanson, A. W., Röhrl, M.: ibid. *B28*, 2032 (1972)
106. Hanson, A. W.: ibid. *B31*, 2352 (1975)
107. Lonsdale, K., Milledge, H. J., Krishna Rao, K. V.: Proc. Roy. Soc. (London), *A255*, 82 (1960)
108. Hope, H., Bernstein, J., Trueblood, K. N.: Acta Crystallogr. *B28*, 1733 (1972)
109. Coulter, C. L., Trueblood, K. N.: ibid. *16*, 667 (1963)
110. Hanson, A. W., Cameron, T. S.: J. Chem. Res. (S) *1980*, 336; ibid. J. Chem. Res. (M) *1980*, 4201
111. Hanson, A. W.: Cryst. Struct. Comm. *9*, 1243 (1980)
112. Hanson, A. W.: Acta Crystallogr. *B28*, 2287 (1972)
113. Hanson, A. W.: ibid. *B33*, 2003 (1977)
114. Umemoto, T., Osubo, T., Misumi, S.: Tetrahedron Letters *1974*, 1573
115. Iwamura, O. H., Kihara, H., Misumi, S., Sakata, Y., Umemoto, T.: Tetrahedron *34*, 3427 (1978)
116. Cf. Keller, H., Krieger, C., Langer, E., Lehner, H.: Monatsh. Chem. *107*, 1281 (1976)
117. Scudder, P. H., Boekelheide, V., Cornutt, D., Hopf, H.: Spectrochimica Acta *37A*, 425 (1981)
118. Ron, A., Schnepp, O.: J. Chem. Phys. *37*, 2540 (1962)
119. Cram, D. J., Allinger, N. L., Steinberg, H.: J. Am. Chem. Soc. *76*, 6132 (1954)
120. Cram, D. J., Bauer, R. H., Allinger, N. L., Reeves, R. A., Wechter, W. J., Heilbronner, E.: J. Am. Chem. Soc. *81*, 5977 (1959)
121. Allinger, N. L., Da Rooge, M. A., Hermann, J. H.: J. Am. Chem. Soc. *83*, 1974 (1961)
122. Ingraham, L. L.: J. Chem. Phys. *18*, 988 (1950)
123. Ingreham, L. L.: ibid. *27*, 1228 (1957)
124. Dewhirst, K. C., Cram, D. J.: J. Am. Chem. Soc. *80*, 3115 (1958)
125. Gleiter, R., Eckert-Makic, M., Schäfer, W., Truesdale, E. A.: Chem. Ber. *115*, 2009 (1982)
126. Truesdale, E. A., Hutton, R. S.: J. Am. Chem. Soc. *101*, 6475 (1979)
127. Burri, K., Jenny, W.: Helv. Chim. Acta *50*, 1978 (1967)
128. Cram, D. J., Dalton, C. K., Knox, G. R.: J. Am. Chem. Soc. *85*, 1088 (1963)
129. Wilson, D. J., Boekelheide, V., Griffin, R. W., Jr.: ibid. *82*, 6302 (1960)
130. Vögtle, F., Neumann, P.: Angew. Chem. *84*, 75 (1972); ibid. Angew. Chem., Int. Ed. Engl. *11*, 73 (1972)
131. Sato, T., Akabori, M., Kainosho, M., Hata, K.: Bull. Chem. Soc. (Japan) *39*, 856 (1966)
132. Sato, T., Akabori, M., Kainosho, M., Hata, K.: ibid. *41*, 218 (1968)
133. Fjukmoto, M., Sato, K., Hata, K.: ibid. *40*, 600 (1967)
134. Kamp, D., Boekelheide, V.: J. Org. Chem. *43*, 3470 (1978)
135. Sherrod, S. A., da Costa, R. L., Barnes, R. A., Boekelheide, V.: J. Am. Chem. Soc. *96*, 1565 (1974)
136. Vögtle, F.: Chem. Ber. *102*, 3077 (1969)
137. Akabori, S., Hayashi, S., Nawa, M., Shiomi, K.: Tetrahedron Letters *1969*, 3727
138. Boekelheide, V., Galuszko, K., Szeto, K. S.: J. Am. Chem. Soc. *96*, 1578 (1974)
139. Fukazawa, Y., Aoyagi, M., Ito, S.: Tetrahedron Letters *1978*, 1067

140. Kawashima, T., Otsubo, T., Sakata, Y., Misumi, S.: ibid. *1978*, 1063
141. Ito, S.: Pure and Applied Chem. *54*, 957 (1982)
142. For an analogous study of the preparation of 2,7-di(*tert*-butyl) derivatives of dihydropyrenes, see Tashiro, M., Takehiko, Y.: J. Am. Chem. Soc. *104*, 3701 (1982) and earlier papers in this series
143. Hanson, A. W.: Acta Crystallogr. *18*, 599 (1965)
144. Hanson, A. W.: ibid. *23*, 476 (1967)
145. Blattmann, H.-R., Schmidt, W.: Tetrahedron *26*, 5885 (1970)
146. Blattman, H.-R., Meuche, D., Heilbronner, E., Molyneaux, R. J., Boekelheide, V.: J. Am. Chem. Soc. *87*, 130 (1965)
147. Schmidt, W.: Tetrahedron Letters *1972*, 581
148. Weaver, L. H., Mathews, B. W.: J. Am. Chem. Soc. *96*, 1581 (1974)
149. Psiorz, M., Hopf, H.: Angew. Chem. *94*, 639 (1982); ibid. Angew. Chem., Int. Ed. Eng. *21*, 623 (1982)
150. (a) Boyd, R. H.: Tetrahedron *22*, 119 (1966);
     (b) Boyd, R. H.: J. Chem. Phys. *49*, 2574 (1968)
151. Shieh, C., McNally, D., Boyd, R. H.: Tetrahedron *25*, 3653 (1969)
152. Lindner, H. J.: ibid. *32*, 753 (1976)
153. Reich, H. J., Cram, D. J.: J. Am. Chem. Soc. *91*, 3517 (1969)
154. Kaup, G., Teufel, E., Hopf, H.: Angew. Chem. *91*, 232 (1979); ibid. Angew. Chem., Int. Ed. Engl. *18*, 215 (1979)
155. Helgeson, R. C., Cram, D. J.: J. Am. Chem. Soc. *88*, 509 (1966)
156. Cram, D. J., Fischer, H. P.: J. Org. Chem. *30*, 1815 (1965)
157. Kleinschroth, J., Hopf, H.: Angew. Chem. *94*, 485 (1982); ibid. Angew. Chem., Int. Ed. Engl. *21*, 469 (1982)
158. Kleinschroth, J., El-tamany, S., Hopf, H., Bruhin, J.: Tetrahedron Letters 3345 (1982)
159. Cram, D. J., Allinger, N. L.: J. Am. Chem. Soc. *77*, 6289 (1955)
160. Jenny, W., Reiner, J.: Chimia *24*, 69 (1970)
161. Marshall, J. L., Folsom, T. K.: Tetrahedron Letters 757 (1971)
162. Marshall, J. L., Song, B.-H.: J. Org. Chem. *39*, 1342 (1974); ibid. *40*, 1942 (1975)
163. Marshall, J. L., Hall, L.: Tetrahedron *37*, 1271 (1981)
164. Reiner, J., Jenny, W.: Helv. Chim. Acta *52*, 1624 (1969)
165. Ciganek, E.: Tetrahedron Letters 3321 (1967)
166. Murad, A. F., Kleinschroth, J., Hopf, H.: Angew. Chem. *92*, 388 (1980); ibid. Angew. Chem., Int. Ed. Engl. *19*, 389 (1980)
167. Garbe, J. E.: Ph. D. Dissertation, Univ. of Oregon 1982
168. Hanson, A. W.: Crystal Structure Comm. *10*, 319 (1981)
169. Hanson, A. W.: ibid. *10*, 559 (1981)
170. Hanson, A. W.: unpublished Work
171. Hanson, A. W.: Crystal Structure Comm. *10*, 195 (1981)
172. Hanson, A. W.: Cryst. Struct. Comm., *11*, 1401 (1982)
173. Erden, I., Gölitz, P., Näder, R., de Meijere, A.: Angew. Chem. *93*, 605 (1981); ibid. Angew. Chem., Int. Ed. Engl. *20*, 583 (1981)
174. Näder, R., de Meijere, A.: Angew. Chem. *88*, 153 (1976); ibid. Angew. Chem., Int. Ed. Engl. *15*, 166 (1976)
175. Horita, H., Otsubo, T., Sakata, Y., Misumi, S.: Tetrahedron Letters 3899 (1976)
176. Menke, K., Hopf, H.: Angew. Chem. *88*, 152 (1976); ibid. Angew. Chem., Int. Ed. Engl. *15*, 165 (1976)
177. O'Connor, J. G., Keehn, P. M.: J. Am. Chem. Soc. *98*, 8446 (1976)
178. Boxberger, M., Volbracht, L., Jones, Jr., M.: Tetrahedron Letters *21*, 3669 (1980)
179. Cram, D. J., Bauer, R. H.: J. Am. Chem. Soc. *81*, 5971 (1959)
180. Staab, H. A., Rebafka, W.: Chem. Ber. *110*, 3333 (1977)
181. Staab, H. A., Herz, C. P., Henke, H.-E.: ibid. *110*, 3351 (1977)
182. Staab, H. A., Haffner, H.: ibid. *110*, 3358 (1977)
183. Staab, H. A., Taglieber, V.: ibid. *110*, 3361 (1977)
184. Schweitzer, D., Hausser, K. H., Taglieber, V., Staab, H. A.: Chem. Phys. *14*, 183 (1976)
185. Vogler, H., Ege, G., Staab, H. A.: Mol. Phys. *33*, 923 (1977)
186. Vogler, H., Ege, G., Staab, H. A.: Tetrahedron *31*, 2441 (1975)

187. Tashiro, M., Koya, K., Yamato, T.: J. Am. Chem. Soc. *104*, 3707 (1982)
188. Laganis, E. D., Finke, R. G., Boekelheide, V.: Proc. Natl. Acad. Aci. USA *78*, 2657 (1981)
189. Cram, D. J., Wilkinson, D. I.: J. Am. Chem. Soc. *82*, 5721 (1960)
190. Langer, E., Lehner, H.: Tetrahedron *29*, 375 (1973)
191. Mourad, A. E., Hopf, H.: Tetrahedron Letters 1209 (1979)
192. Cristiani, F., DeFilippo, D., Deplano, P., Devillanova, F., Diaz, A., Trogu, E. F., Verani, G.: Inorg. Chim. Acta *12*, 119 (1975)
193. Ohno, H., Horita, H., Otsubo, T., Sakata, Y., Misumi, S.: Tetrahedron Letters 265 (1977)
194. Elschenbroich, C., Möckel, R., Zennek, U.: Angew. Chem. *90*, 560 (1978); ibid. Chem. Int. Ed. Engl. *17*, 531 (1978)
195. Bennett, M. A., Matheson, T. W.: J. Organomet. Chem. *175*, 87 (1979)
196. Bennett, M. A., Matheson, T. W., Robertson, G. B., Smith, A. K., Tucker, P. A.: Inorg. Chem. *19*, 1014 (1980)
197. Laganis, E. D., Finke, R. G., Boekelheide, V.: Tetrahedron Letters *21*, 4405 (1980)
198. Laganis, E. D., Voegeli, R. H., Swann, R. T., Finke, R. G., Hopf, H., Boekelheide, V.: Organometallics *1*, 1415 (1982)
199. Swann, R. T.: Ph. D. Dissertation, Univ. of Oregon 1983
200. Gill, T. P., Mann, K. R.: Organometallics *1*, 485 (1982)
201. Rohrbach, W., Boekelheide, V.: Unpublished Work
202. Gill, T. P., Mann, K. R.: Inorg. Chem. *19*, 3007 (1980)
203. Gill, T. P., Mann, K. R.: J. Organomet. Chem. *216*, 65 (1981)
204. Nesmeyanov, A. N., Vol'kenau, N. A., Bolesova, I. N.: Tetrahedron Letters 1725 (1963)
205. Lee, C. C., Steele, B. R., Sutherland, R. G.: J. Organomet. Chem. *186*, 265 (1980)
206. Koray, A. R.: ibid. *212*, 233 (1981)
207. Elzinga, J., Rosenblum, M.: Tetrahedron Letters *23*, 1535 (1982)
208. Swann, R. T., Boekelheide, V.: J. Organomet. Chem. *231*, 143 (1982)
209. Langer, E., Lehner, H.: ibid. *173*, 47 (1979)
210. Finke, R. G., Voegeli, R. H., Lagenis, E. D., Boekelheide, V.: Organometallics, *2*, 347 (1983)
211. Darensbourg, M. Y., Muetterties, E. L.: J. Am. Chem. Soc. *100*, 7425 (1978)
212. Huttner, G., Lange, S.: Acta Crystall. *B28*, 2049 (1972)
213. Geiger, Jr., W. E., Bowyer, W., Boekelheide, V.: Unpublished Work
214. Hamon, J.-R., Astruc, D., Michaud, P.: J. Am. Chem. Soc. *103*, 758 (1981), and earlier papers this series
215. Vögtle, F., Klieser, B.: Synthesis *1982*, 294
216. Klieser, B., Vögtle, F.: Angew. Chem. *94*, 632 (1982); ibid, Angew. Chem., Int. Ed. Engl. *21*, *21*, 618 (1982)
217. Klieser, B., Vögtle, F.: Angew. Chem. *94*, 922 (1982); ibid, Angew. Chem., Int. Ed. Engl. *21*, 928 (1982)
218. El-tamany, S., Hopf, H.: Chem. Ber., *116*, 1682 (1983)

Water Soluble Cyclophanes as Hosts and Catalysts

Iwao Tabushi and Kazuo Yamamura

Department of Synthetic Chemistry, Kyoto University, Japan

Table of Contents

Iwao Tabushi and Kazuo Yamamura

Progress of this field of chemistry in the last ten years is reviewed on synthesis, static and dynamic molecular structures of macrocyclophanes, guest accomodation in solution and in crystalline phase, and some catalytic reactions on accomodated substrates. Hydrophobic guest recognition — the most fundamental aspect in guest binding in water — by synthetic hosts is discussed to show their molecular aspects or energetics, quoting some relevant studies on cyclodextrin inclusions. Accumulated spectroscopic data support that the one to one specific inclusion is achieved by cyclophanes. Some interesting informations as to inclusion geometry are also provided by X-ray and NMR.

Abbreviations

$\overbrace{}^{n}$

$n°$-PCP [2.2.2 ⋯ 2]paracyclophane (abbreviated also as [2^n]PCP)

N_4PCP N,N′,N″,N‴-tetramethyl-2,11,20,29-tetraaza[3.3.3.3]paracyclophane

N_6PCP N,N′,N″,N‴,N⁗,N⁗′-hexamethyl-2,11,20,29,38,47-hexaaza[3.3.3.3.3.3]-paracyclophane

N_4^+PCP N,N,N′,N′,N″,N″,N‴,N‴-octamethyl-2,11,20,29-tetraazonia[3.3.3.3]para-cyclophane tetrakistetrafluoroborate

S_2PCP 2,11-dithia[3.3]paracyclophane

S_4PCP 2,11,20,29-tetrathia[3.3.3.3]paracyclophane

S_2^+PCP S,S′-dimethyl-2,11-dithionia[3.3]paracyclophane bistetrafluoroborate

S_4^+PCP S,S′,S″,S‴-tetramethyl-2,11,20,29-tetrathionia[3.3.3.3]paracyclophane tetrakistetrafluoroborate

A Introduction

"Host-guest chemistry"[1] is an exciting and rapidly growing field of organic chemistry dealing with design, synthesis, and application of artificial organic compounds capable of "molecular recognition". A very promising application of this field of research involves the preparation of an artificial specific receptor, transport carrier, or potent catalyst, i.e., a molecule or molecular system that by its dramatic "extra" function achieves very specific molecular recognition. Molecular recognition, a relatively new concept to guide this new chemistry, has been significantly pursued and developed for the last decade with cyclodextrins[2-6], crown ethers[7], and cryptands[8]. More recently, however, another unique class of compounds, *macrocyclic cyclophanes*, has been developed so that the host-guest frontier has been extensively advanced to a unique area of *water soluble cyclophanes as hosts and catalysts*.

Completely artificial host molecules of the cyclophane type have several advantages a) straightforward preparation, b) the well-defined molecular dimension, size or shape (very often allowing X-ray crystallographic study), c) the easy-to-get information as to various physicochemical properties such as macroring conformation and internal rotation of aromatic rings, and d) remarkable thermal and/or chemical stability. Moreover, a considerable variation of cyclophane's structure makes it easy to prepare *modified* macrocyclic host molecules for very specific interaction with certain guest molecules. This article is mainly concerned with the progress of cyclophane chemistry in the host-guest field in the past decade; several other reviews on cyclodextrin, crown ether and cryptand add mutually complementary arguments to this review.

The inclusion phenomenon by a macrocyclic cyclophane was first observed by Stetter and Roos in 1955; they reported that the cyclic tetraamine *1b* (n = 3) or *1c* (n = 4) forms a stable 1:1 complex with benzene or dioxane, whereas no similar complex formation was observed for *1a* (n = 2) of smaller ring size[9].

$1a-c$

(n = 2, 3, 4)

This discovery was followed by many synthetic and structural studies of macrocyclic cyclophanes by Stetter[10-15] and others[16-23]. More recently, a unique type of cavities containing macrocycles was developed by Vögtle[24-26]. With some macrocyclic cyclophanes such as cyclotrivelatrylene[27-30] or metacyclophanes[31-33], cage or channel inclusions are formed. In this review, these are classified as lattice (cavity) inclusions and are differentiated from *molecular* cavity inclusion compounds discussed in Chapter D.IV.

Crystal lattice inclusion:

$$n \text{ Host } + m \text{ Guest } \rightarrow (\text{Host}_n \text{ Guest}_m)$$
$$\text{lattice inclusion, channel crystal}$$

Molecular cavity inclusion:

$$n \text{ Host } + m \text{ Guest } \rightarrow (\text{Host Guest})_n$$
$$\text{host-guest molecular inclusion crystal}$$

Lattice inclusion usually gives a host:guest ratio of 1:n and guest selectivity may be poor. As to molecular cavity inclusion, no satisfactory example has been provided until recently when the bioorganic chemistry of water soluble cyclophanes was studied.

For modelling hydrophobic enzyme-substrate or receptor-substrate complexation, Tabushi started the preparation of water soluble cyclophanes. A generalized feature of host-guest chemistry in aqueous solution mostly established for cyclodextrins was thus further extended to cyclophane type hosts, revealing many fundamental aspects of cyclophane's binding and catalysis e.g., molecular dynamics of inclusion, structural information of host molecule, or geometrical optimization of host-guest recognition. Later, Murakami also worked out [20]paracyclophane and its derivatives for enzyme-like catalysis, which is included in another chapter of this book. More recently, Koga et al. and Whitlock et al. devised other types of water soluble cyclophanes. This article attempts to briefly review these subjects.

B Water Soluble Cyclophanes

Molecular recognition is ubiquitous in nature. For example, an enzyme protein recognizes molecular shape and size of substrate or inhibitor very precisely (Table I). It is well known that even a hydrolytic enzyme chymotrypsin, one of the simplest enzymes in living systems, shows a remarkably high specificity toward amino acid residues having a hydrophobic side chain. Interestingly, the high specificity is not restricted to natural side chains — indolyl, phenyl, p-hydroxyphenyl — but also non-natural side chains such as cyclohexyl (see Table I). Most importantly, the free energy change amounts to 0.6–0.7 kcal/mol per CH_2 or CH unit for these substrates, being in good agreement with a hydrophobic interaction free energy change of 0.6–0.9 kcal/mol.

Water molecules tend to shift to more structured states along a hydrophobic surface in water according to Nemethy and Scheraga [34]. Such a water assembly (Fig. 1.) is only formed by sacrificing considerable motional freedom of water molecules, leading to a large entropy decrease. From this reason, hydrophobic molecules in water tend to associate in order to reduce the hydrophobic surface area that is originally exposed to water. Fig. 1 is a supersimplified picture of the "hydrophobic interaction". This argument leads to an important consequence —

Table I. Association Constants for Binding of Specific Substrates to Several Enzymes

Enzyme	Substrate	$1/K_s$ or $1/K_m$ (M^{-1})	Ref.

Carboxypeptidase	$Cl-\phi-CH=CH-CO_2-CH-CO_2H$
	              $|$
	              $CH_2\phi$
		740	[64a]

Lisozyme	$(NAG)_3$	12 500	[64b]
Carbonic Anhydrase	$C_6H_5SO_2NH_2$	600 000	[64c]

Chymotrypsin	$AcNH-CH-CO_2Et$
	           $|$
	           R
			[64d]

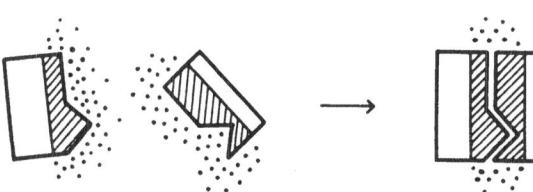

R = HO—⟨C₆H₄⟩—CH₂—   1 400

⟨C₆H₅⟩—CH₂—   800

indolyl-CH₂—   10 000

cyclohexyl-CH₂—   5 000
CH₃—   2
H—   10

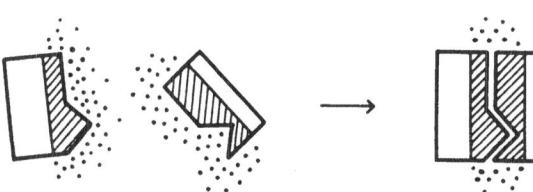

Fig. 1. Schematic representation of disappearance of more structured water region (dotted area) along hydrophobic surface (hatched area) by hydrophobic interaction

a cyclophane molecule having on appropriate molecular cavity, once solubilized in water, must provide a strong, hydrophobic domain — the first requisite for hydrophobic host-guest recognition.

To solubilize a cyclophane molecule in water,

a) a C_nH_m molecule with n < 10 is combined with a strong, polar (hydrophilic) group like N^+, S^+, SO_3^-, (such as $Et_4N^+X^-$);

b) a more hydrophobic C_nH_m molecule with n > 16 usually self-aggregates to form micelles, vesicles, or higher aggregates depending upon the nature and number of polar groups and the nature of the C_nH_m moiety;

c) a molecule of moderate hydrophilicity-hydrophobicity shows phase-transfer characteristics.

The argument for the hydrophobic-hydrophilic balance puts forth the structural requirement for water soluble cyclophanes. In order to meet the requirements, the following general strategies of the host design are applied:

a) Aromatic rings are joined, (Fig. 2a) with very hydrophilic groups of smallest size. Hydrophilic segments can be NR_4^+, SR_3^+, PR_4^+, or other ionic groups.

b) Aromatic groups may be replaced by suitable rigid segments (e.g., $-C\equiv C-$, $-CH=CH-$, etc.) (Fig. 2b) to afford a water soluble macrocycle having a somewhat shallower hydrophobic cavity.

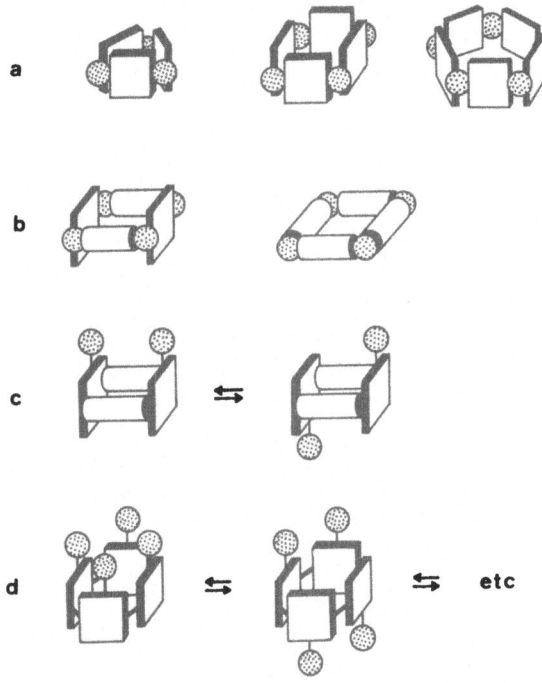

Fig. 2. Strategy for constructing water soluble macrocyclic cyclophanes that are composed of aromatic rings (square plate), hydrophobic rigid segment (rod), and hydrophilic ionic group (dotted circle)

c) Ionic groups can be appended to aromatic groups (Fig. 2c). The diacetylenic group $-C\equiv C-C\equiv C-$ was successfully used as the rod portion to connect two benzene rings or two naphthalene rings as in the compound *17* or *20*.

d) Macrocyclic cyclophanes like higher [2ⁿ]paracyclophanes are easily made water soluble by introducing appropriate hydrophilic groups onto aromatic nuclei e.g., via electrophilic substitution reactions (Fig. 2d).

The cyclophane structure — ring size, cavity shape, cavity depth, and functionality — is easily varied; in Eqs. (1)–(14) water soluble cyclophanes are shown together with their preparation procedures.

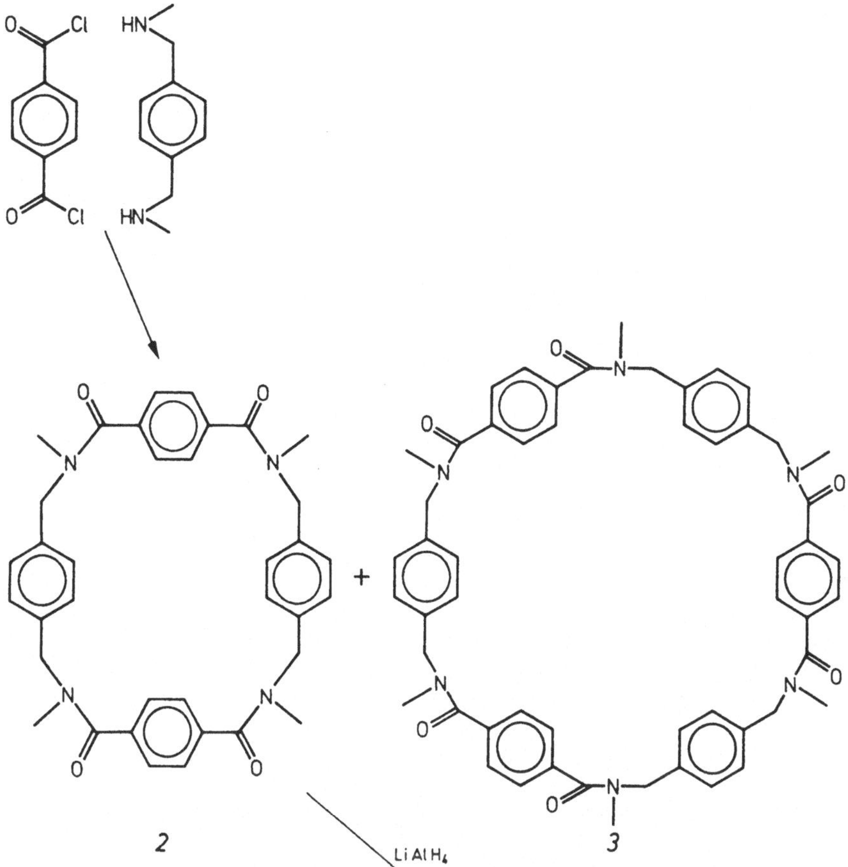

$$(1)$$

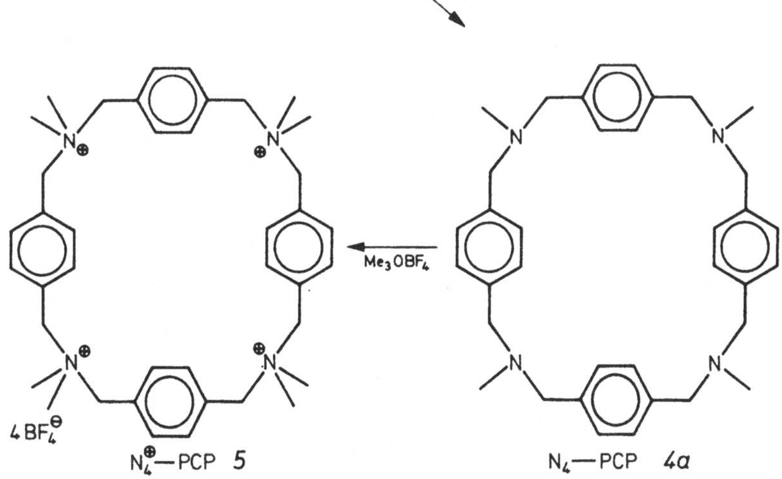

$$(2)$$

$$4b \quad N_6-PCP \xleftarrow[\text{LiAlH}_4]{} 3 \qquad (3)$$

151

B.I Halide-Amine or Halide-Thiol Cyclizations

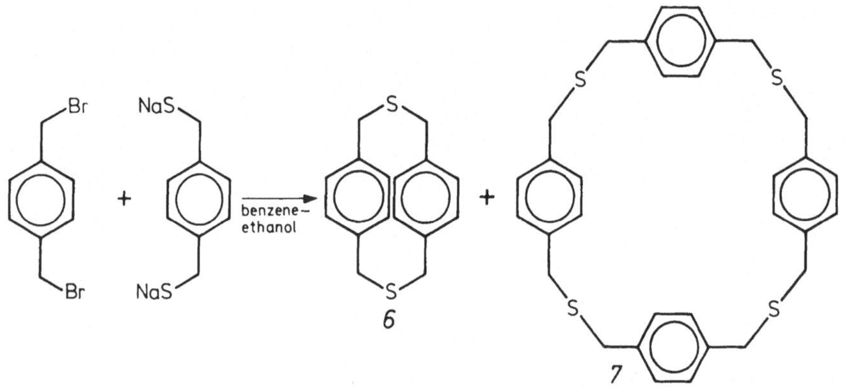

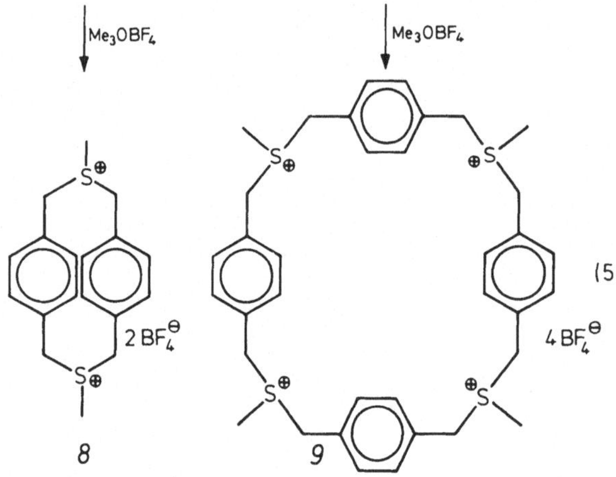

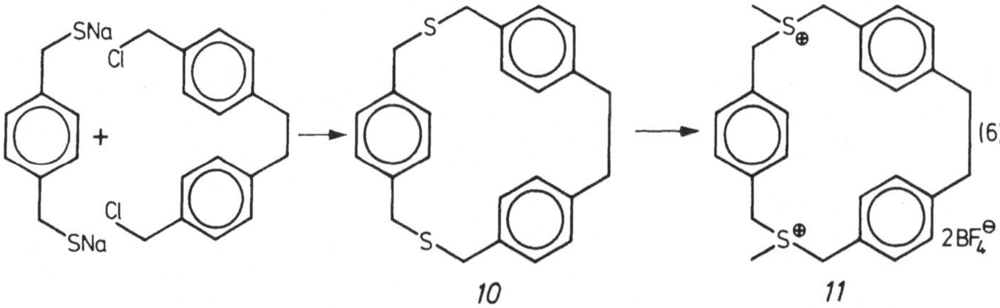

The reaction most frequently utilized for preparation of water soluble cyclophane is one-stage halide-amine or halide-thiol cyclization under *high dilution* (see contribution by Vögtle et al. in this volume). N,N′,N″,N‴-Tetramethyl-1,12,19,30-tetraoxo-2,11,20,29-tetraaza[3.3.3.3]paracyclophane (*2*), for example, is prepared in a fairly good yield (10.5%) from terephthaloyl chloride and N,N′-dimethyl p-xylylene diamine [35, 36–38]. Reduction of *2* with LiAlH$_4$ in dioxane gives a water soluble cyclic tetraamine N,N', N",N'''-tetramethyl-2,11,20,29-tetraaza[3.3.3.3]paracyclophane (*4a*) in 52% yield, and *4a* is permethylated with Meerwein's reagent affording the tetra-ammonium compound *5* in 40% yield after recrystallization [37, 38]. Compound *4* is soluble in water below pH 6 and compound *5* is more soluble over a wide pH range from 4 to 13.

The halide-thiol cyclization is also recommendable for macrocycle preparation as far as the cyclization yield is concerned; e.g., 2,11,20,29-tetrathia[3.3.3.3]para-cyclophane (*7*) is obtained in a good yield of 20% together with the cyclic dimer 2,11-dithia[3.3]paracyclophane (*6*) (31%) from p-xylylene dibromide and p-xylylene dithiol [39].

Similarly, 2,11-dithia[3.3.2]paracyclophane (*10*) is prepared by using halide-thiol cyclization as shown in Eq. (6) [40]. Each of the cyclic sulfides *6*, *7* or *10* is converted to the corresponding sulfonium compounds *8*, *9* or *11*, respectively. Those sulfonium salts were used as host molecules in water in the inclusion study by Tabushi et al. [41].

As a very promising host that can be readily "functionalized" with appropriate functional groups, the authors have recently exploited 2,11-dimethyl-2,11,20,29-

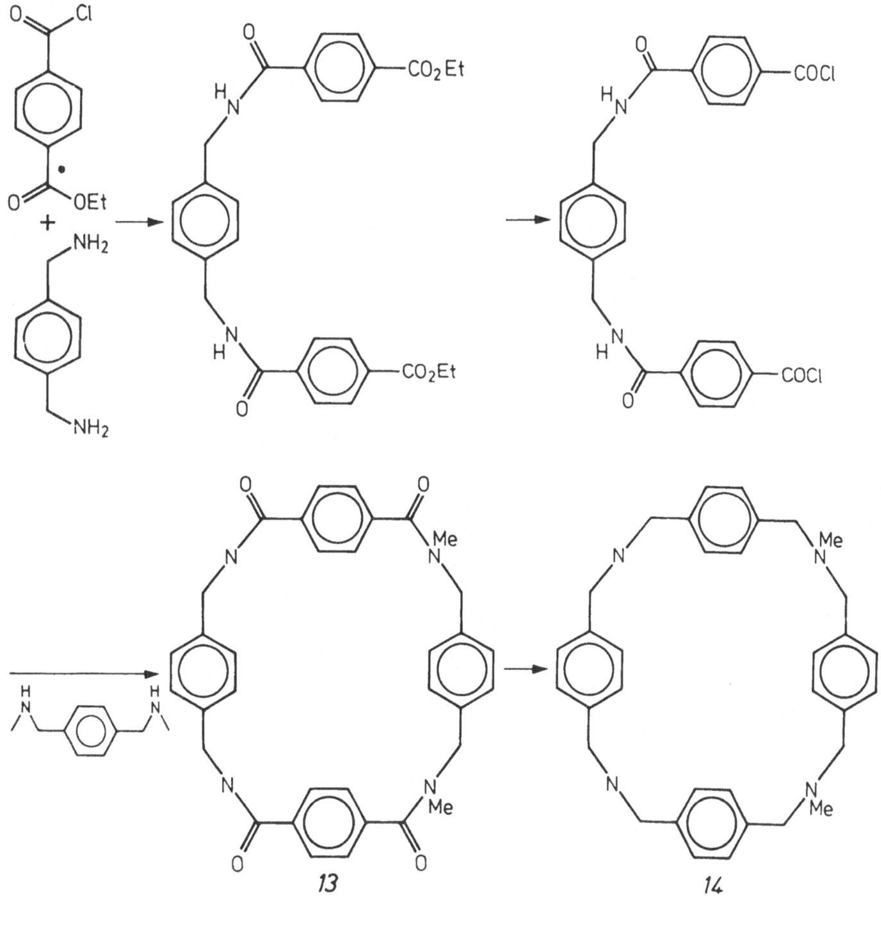

$$ (7) $$

13   *14*

tetraaza[3.3.3.3]paracyclophane (14) bearing two secondary N—H groupings [41]. Compound 14 is prepared stepwise via halide-amine reaction as shown in Eq. (7), and 14 was successfully "functionalized" with a hydroxamate grouping known as a strong catalytic group for the ester hydrolysis.

Koga et al. applied the N-tosylamide-halide reaction for the preparation of 1,6,20,25-tetraaza[6.1.6.1]paracyclophane tetratosylate (12b) which was detosylated to give the parent cyclic tetraamine (12a) [Eq. (8)] [42].

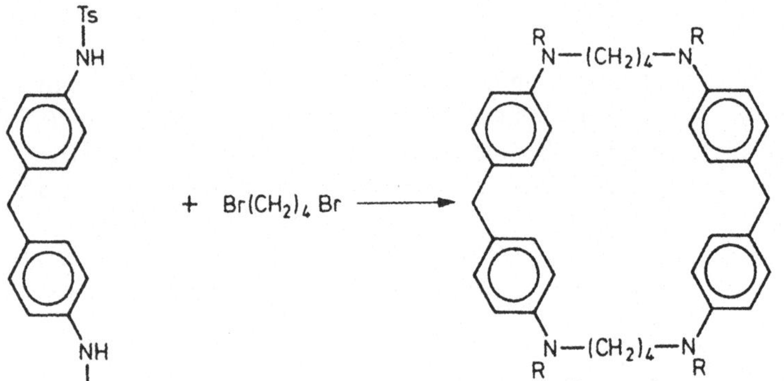

$$12a \quad R = H$$
$$b \quad R = Ts$$
$$c \quad R = Ac$$

(8)

B.II Cyclization via the Acetylenic Coupling Reaction

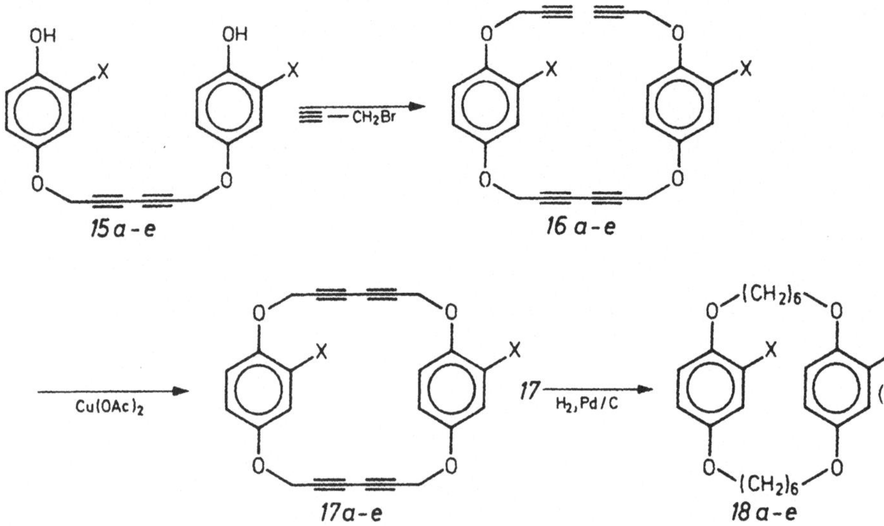

15 a - e   16 a - e

17a-e   18 a - e

(9)

a X = CO$_2$H
b X = CO$_2$CH$_3$
c X = CO$_2$CH$_2$CCl$_3$
d X = CO$_2$C$_6$H$_{13}$
e X = CO$_2$K

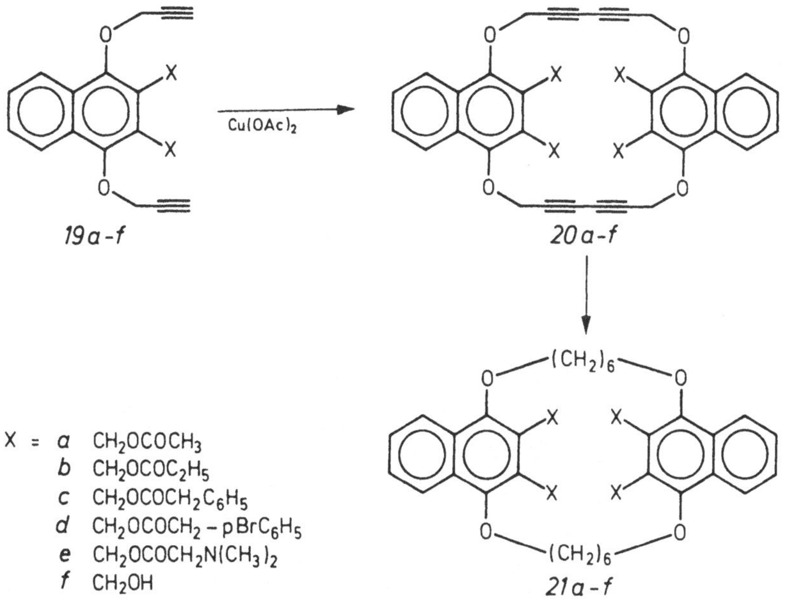

$$19a-f \xrightarrow{Cu(OAc)_2} 20a-f \qquad (10)$$

X = a CH_2OCOCH_3
b $CH_2OCOC_2H_5$
c $CH_2OCOCH_2C_6H_5$
d $CH_2OCOCH_2 - pBrC_6H_5$
e $CH_2OCOCH_2N(CH_3)_2$
f CH_2OH

21a-f

The oxidative coupling of acetylene group with cupric acetate in pyridine was used by Whitlock et al. to prepare 1,6-hexadi-2,4-ynyl ether linked paracyclophanes 17a–e and 1,4-naphthalenophanes 20a–f [Eqs. (9) and (10)] [43, 44]. The cyclization proceeds smoothly at ordinary concentration (67% cyclization yield of 17b from 23 mM 16b) and high-dilution is not necessary. Their stepwise coupling, from 2-hydroxy-5-propargyloxybenzoate to 15 and from 16 to 17, was unsuccessful for 1,4-naphthaleno-phanes 20, but instead the direct cyclodimerization of 19 gave 20 in 35–40% yield, except 20e. Compound 20e is prepared from 20a by methanolysis with K_2CO_3 followed by acylation of 20f with N,N-dimethylglycine hydrochloride. Alkaline hydrolysis is successfully carried out on n-hexyl ester 17d but not on methyl ester 17b due to its insolubility in conventional solvents.

B.III Oligocyclization of p-Xylylene Chloride by Modified Wurtz Coupling Reactions

The Wurtz reaction is of considerable utility in the preparation of cyclophanes and, in particular, arylmethyl halides give satisfactory results. In an attempt to prepare [2.2]paracyclophane from p-xylylene bromide and sodium metal, Baker et al. isolated the trimer, [2.2.2]paracyclophane 22a in 4% yield [45]. More recently, Tabushi et al. successfully prepared a series of higher members of $[2^n]$paracyclophanes, viz., [2.2.2]paracyclophane (22a, 3°-PCP) [46], [2.2.2.2]paracyclophane (22b, 4°-PCP) [47], [2.2.2.2.2]paracyclophane (22c, 5°-PCP) [48], [2.2.2.2.2.2]paracyclophane (22d, 6°-PCP) [48], and [2.2.2.2.2.2.2]paracyclophane (22e, 8°-PCP) [48] from p-xylylene chloride by using sodium tetraphenylethylene. This new preparation is a remarkable improvement of the old experiments carried out by Müller and Röscheisen who obtained only polymer with this modified Wurtz coupling of p-xylylene bromide [49].

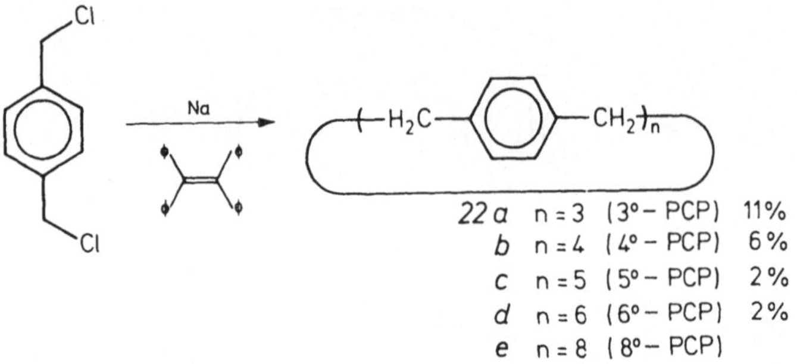

Cl
Na
H2C
CH2
n
Cl
22a n = 3 (3°- PCP) 11%
b n = 4 (4°- PCP) 6%
c n = 5 (5°- PCP) 2%
d n = 6 (6°- PCP) 2%
e n = 8 (8°- PCP)

(11)

The modified Wurtz cyclization has a greater synthetic advantage for cyclic oligomers of p-xylylene, since the pyrolysis of p-xylene affords 3°-PCP (*22a*) [50] and 4°-PCP (*22b*) [51] in trace amounts. Errede et al. reported a yield less than 0.05% for 4°-PCP from p-xylene [51].

Furthermore, 1,2-bis(4-chloromethylphenyl)ethane (*23*) can be a precursor for the preparation of 4°-PCP, 6°-PCP, and 8°-PCP [48].

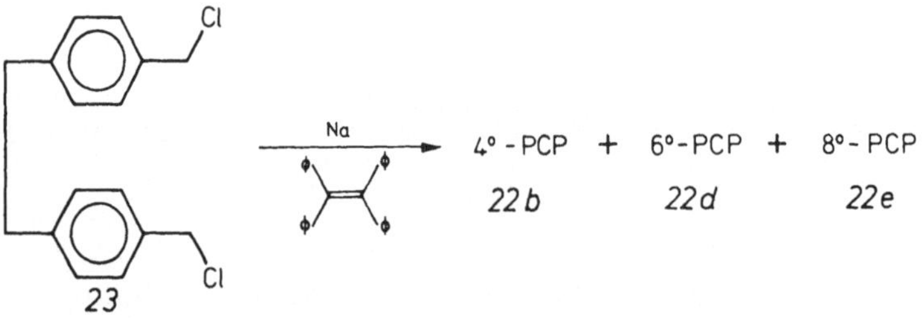

Cl
Na
4° - PCP + 6°-PCP + 8°- PCP
22b
22d
22e
Cl
23

(12)

B.VI Modification by Electrophilic Substitution

Equations (13)–(14) show several substitution reactions of n°-PCPs that were used to prepare water soluble cyclophanes *24*, *25*, *30*, where the conventional electrophilic reactions on aromatic nuclei are successfully applied in reasonable yields [52, 53].

5°-PCP
or
6°-PCP
1.) ClCH2OCH3
ZnCl2
2.) N(CH3)3
X
X= CH2N(CH3)3
24
25

(13)

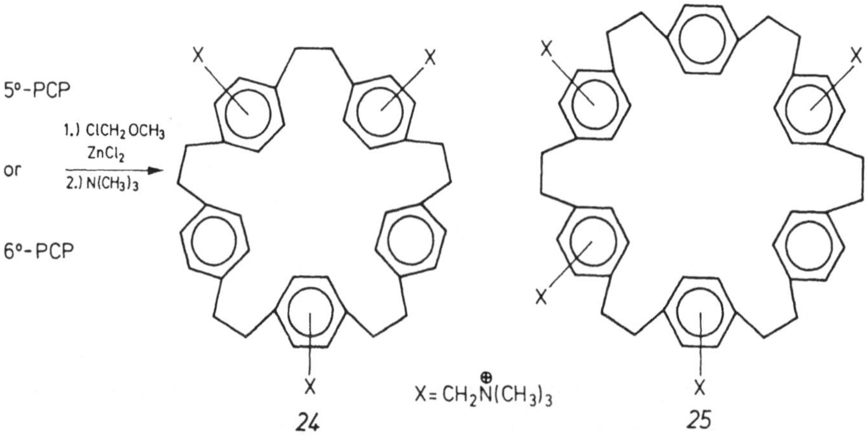

$$4^\circ\text{-PCP} \xrightarrow{N_2O_5} 4^\circ\text{-PCP-NO}_2 \xrightarrow[\text{AlCl}_3]{\text{Ac Cl}} \underset{\underset{\text{Ac } 27\,(86\%)}{|}}{4^\circ\text{-PCP-NO}_2} \xrightarrow{\text{KOBr}}_{\text{CH}_2\text{N}_2} \underset{\underset{28\,(94\%)}{\overset{|}{\text{CO}_2\text{CH}_3}}}{4^\circ\text{-PCP-NO}_2} \xrightarrow{\text{Pd}/\text{C}} \underset{\underset{29\,(92\%)}{\overset{|}{\text{CO}_2\text{CH}_3}}}{4^\circ\text{-PCP-NH}_2}$$

$$26\,(67\%)$$

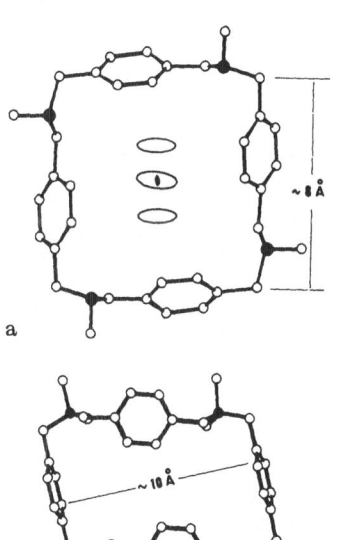

$$\xrightarrow{\quad} \underset{\underset{30\,(45\%)}{\overset{|}{\text{CO}_2\text{CH}_3}}}{4^\circ\text{-PCP-NH-CH}_2}$$

(14)

C Molecular Structure of Water Soluble Cyclophanes

Water soluble cyclophanes are designed not only to provide a strong, hydrophobic cavity but also to adjust the macroring's shape most appropriate for inclusion of certain guests' shapes. This is demonstrated by X-ray crystallography with the $N_4PCP-CHCl_3$ inclusion complex and N_4^+PCP (5) as shown in Fig. 3.

a

b

Fig. 3a and b. Crystal structures of **a** N,N',N'',N'''-tetra-methyl-2,11,20,29-tetraaza[3.3.3.3]paracyclophane (4a) · CHCl₃ complex and **b** N,N,N',N',N'',N'',N''',N'''-octa-methyl-2,11,20,29-tetraaza[3.3.3.3]paracyclophane tetraammonium (5)

N_4PCP (4a) has an approximately square shape and N_4^+PCP (5) has a rhomboidal shape. The shape of the macroring is mostly determined by the connecting segment $-CH_2-N(R)-CH_2-$ which resists to take up an unstable eclipsed configuration but favors $(-)$gauche-trans/gauche-trans configurations in N_4PCP (4a) or gauche-$(-)$trans/trans-$(-)$gauche configurations in N_4^+PCP (5). Thus, N_4PCP (4a) provides a square cavity surrounded by four benzene "walls", and chloroform (guest) is

157

Table II. UV Spectral Data[a] of Higher [2^n]Paracyclophanes (22)

Compound	λ_{max}, nm (log ε)			
2°-PCP[b]	302$_{sh}$ (2.19)		285 (2.41)	
3°-PCP	276 (2.89)	269$_{sh}$ (2.88)	267 (2.95)	262 (2.83)
4°-PCP	274 (3.13)	267$_{sh}$ (3.16)	265 (3.20)	260 (3.09)
5°-PCP	274 (3.24)	267.5$_{sh}$ (3.22)	265.5 (3.28)	260 (3.15)
6°-PCP	273.5	267$_{sh}$	265	260
8°-PCP	274	267.5$_{sh}$	265.5	260
p-Xylene[c]	274 (2.85)	269 (2.75)	266 (2.73)	260 (2.60)
N_4^+ PCP (5)[d]	274	268 (3.4)	262$_{sh}$	

[a] In n-hexane
[b] Cram, D. J. and Steinferg, H., J. Am. Chem. Soc. 73, 5691 (1951)
[c] In n-heptane
[d] In 1/15 M phosphate buffer, pH 7

Table III. ^{1}H-NMR Spectra of Water Soluble Heterocyclophane

	δ_{arom}	δ_{CH_2}	δ_{Me}
N$_4$PCP (4a)	7.23	3.33	2.33
N$_6$PCP (4b)	7.27	3.47	2.17
S$_2$PCP (6)			
S$_4$ PCP (7)	7.18	3.55	—

included almost "inside" the cavity with nearly maximum van der Waals stabilization (see Sect. D.IV) [54]).

The benzene rings in 4a are normal (undistorted) according to their electronic spectra (Table II) and seem to keep freedom of internal rotation (NMR spectra in Table III). Interestingly, an "all face" conformation (Fig. 3.) is favored with N$_4$PCP (4a) as found for the parent cyclophane, tetrakis[2.2.2.2]paracyclophane (22b)

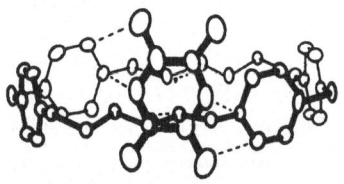

Fig. 4. Crystal structure of a durene complex of 1,6,20,25-tetraaza[6.1.6.1]paracyclophane · 4 HCl

in solution by NMR (Sect. D.I). In the crystalline phase, however, the lattice interaction between N_4PCP (*4a*) molecules as well as host-guest molecular interaction seems very important (Sect. D.IV).

It is most plausible that the two "face"-two "lateral" structure of N_4^+PCP (*5*) in the crystalline state changes to "all face", when it accomodates such hydrophobic guests as benzene or naphthalene (cf. Figs. 4 and 5). For the inclusion, a ring size of ca. 8 Å for *4a* or of ca. 10 Å for *5* is the necessary and sufficient size to give 1 to 1 host-guest molecular inclusion with a guest molecule of the benzene or naphthalene type.

In conclusion, cyclophanes prove to be excellent ring systems for inclusion of hydrophobic organic guest molecules [77], another example is a durene complex of 1,6,20,25-tetraaza[6.1.6.1]paracyclophane (*12a*) · 4 HCl reported by Koga et al. [42]. The cavity, which has "rectangularly shaped open-ends ($\sim 3.5 \times 7.9$ Å) and a depth of 6.5 Å", includes durene as a guest molecule as shown in Fig. 4.

D Host-Guest Binding

D.I Face Conformation of Higher n°-Paracyclophanes

Another piece of evidence for *cavity* inclusion is obtained by NMR, i.e., the time-averaged conformational picture exhibited by higher n°-PCP's is approximated by a "face" conformation rather than a "lateral" conformation as shown in Fig. 5, based on the following results:

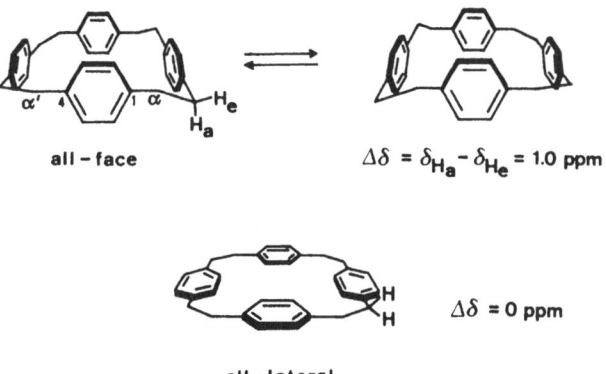

all – face

$\Delta\delta = \delta_{H_a} - \delta_{H_e} = 1.0$ ppm

$\Delta\delta = 0$ ppm

all – lateral

Fig. 5. All face conformation of tetrakis[2.2.2.2]paracyclophane and hypothetical all lateral conformation

Each of the n°-PCP's (n = 3–6, 8) (*22*) shows two singlet ^{1}H NMR absorptions due to aromatic and aliphatic protons with the following δ values [48]:

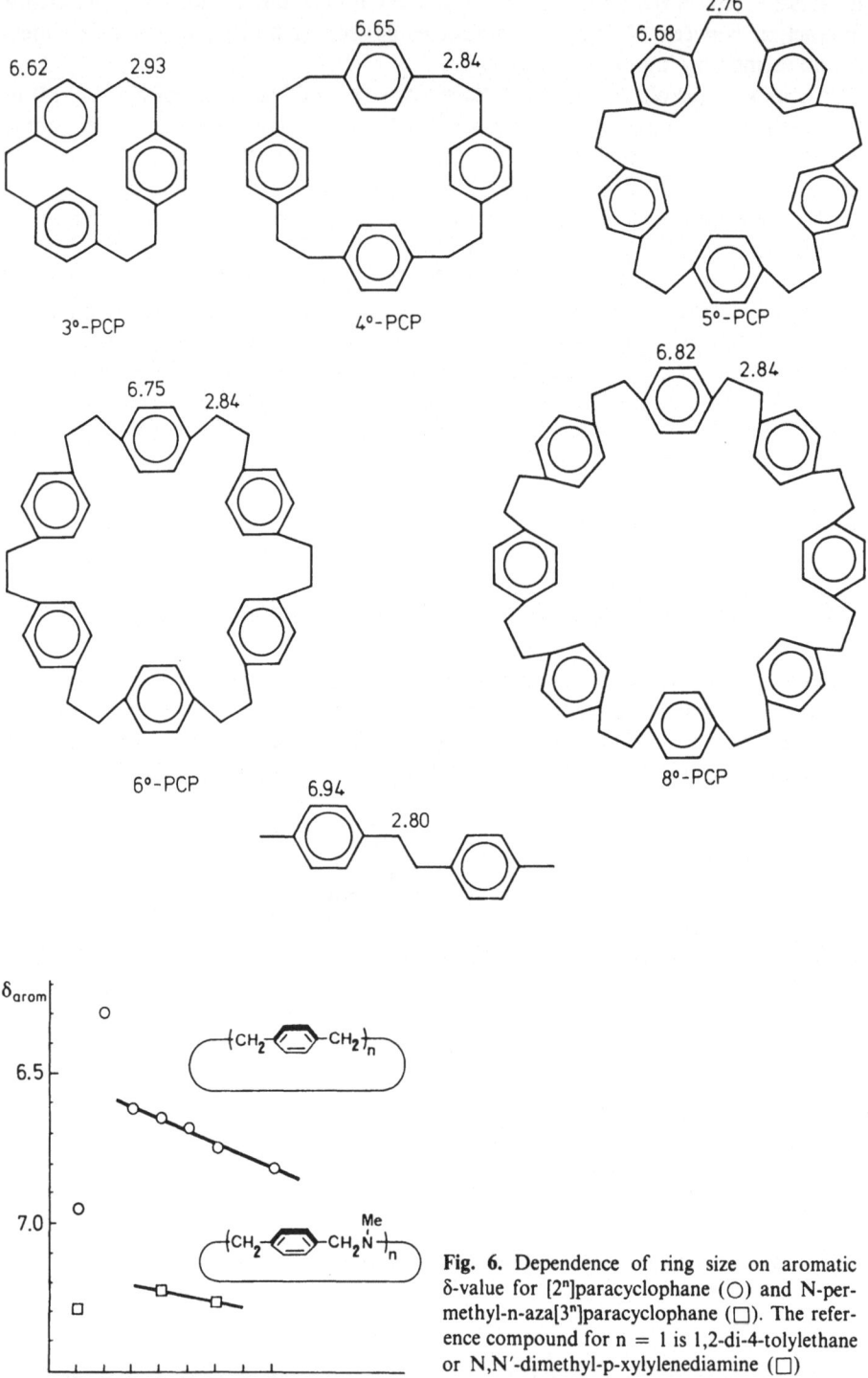

Fig. 6. Dependence of ring size on aromatic δ-value for [2^n]paracyclophane (○) and N-permethyl-n-aza[3^n]paracyclophane (□). The reference compound for n = 1 is 1,2-di-4-tolylethane or N,N′-dimethyl-p-xylylenediamine (□)

Clearly, the aromatic δ values show a considerable shielding effect, the magnitude of which decreases with increase of the ring size (Fig. 6). The shielding effect is ascertained by independent results: Substituents in 4-substituted 4°-PCP turn outside, resulting in a much larger shielding effect (of π-ring current of other "face" benzenes) on aromatic δ value of the substituted benzene (Sect. D.II).

The ethylene protons of 4°-PCP (*22b*) coalesce at −71 °C in $CDCl_3-CH_2Cl_2$ or −85 °C in CS_2 ($\Delta G^{\ddagger}_{-85°} = 9.1$ kcal/mol) below which they appear as a AB quartet (H_a, H_e) with a chemical shift difference, $\Delta\delta = \delta_{H_a} - \delta_{H_b} = 0.5$ ppm (Fig. 7.). The

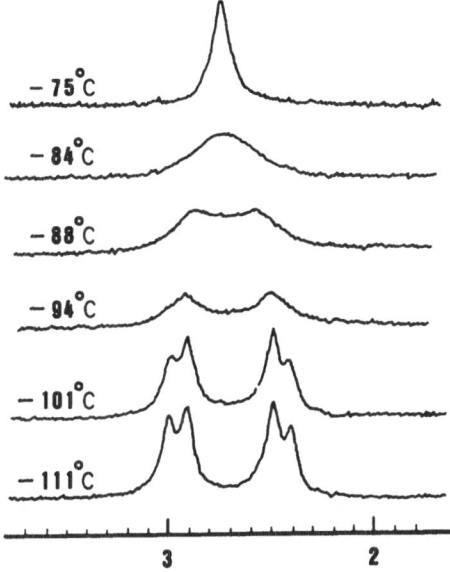

Fig. 7. ^{1}H-NMR absorption of methylene protons of tetrakis[2.2.2.2]paracyclophane (4°-PCP) at low temperatures. 100 MHz, CS_2

chemical shift difference of 0.51 ppm (limiting value) observed for the two frozen protons is ascribed to the axial-equatorial difference (Fig. 5.). based on theoretical calculation of the shielding effect (Johnson's equation) viz., $\Delta\delta_{calcd} = 1.0$ ppm for the "all face" and 0.0 ppm for the "all lateral" conformation. The observed magnitude, 0.51 ppm, suggests that four benzene rings still vibrate or rotate (around a $C_\alpha-C_1-C_4-C_{\alpha'}$ axis), in agreement with only a little line broadening observed for the aromatic proton absorptions (Table IV) [48]. (See also ref. [85])

The unexpectedly small $\Delta G^{\ddagger}$ value of 3°-PCP is ascribed to its slightly repulsive [55] transannular distance of benzenes. This forces the benzene rings to be apart,

Table IV. Line Broadening[a] of n°-PCP at −75° in $CDCl_3-CH_2Cl_2$

Proton	3°-PCP	4°-PCP	5°-PCP	6°-PCP
CH_2	4.9/2.0	[b]	7.5/1.8	7.5/3.3
Aromatic	3.1/2.0	2.8/1.8	2.5/1.8	3.0/3.3

[a] Relative half-width in Hz compared with that of added CH_2Cl_2 (standard)
[b] Below the coalescence temp. see Fig. 7

making the conformation nearly eclipsed and raising the energy of the gauche conformation (Fig. 8.).

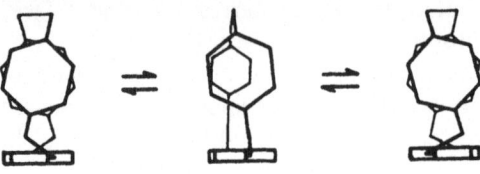

Fig. 8. Conformation change (pseudo rotation) in 3°-PCP

Direct experimental evidence for the $p\pi$-$p\pi$ interaction is
a) a considerable hypochromic effect in the electronic spectrum of 3°-PCP (Table II) [55];
·b) a considerable transannular control of the acetylation rate.

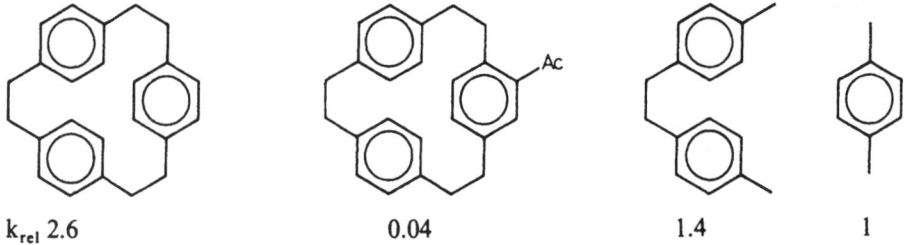

k_{rel} 2.6   0.04   1.4   1

Relative Rate of Acetylation, at 2° in CH_2Cl_2

Thus, the small rate enhancement in the first acetylation is attributable to "transannular" stabilization of the benzene-$Ac^+\sigma$ complex (*31a*) by other benzene rings (*31b* and *c*), while the marked rate retardation (1/63 times) in the second (and third)

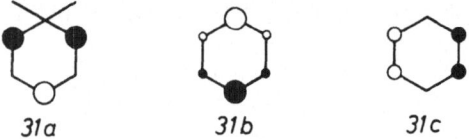

31a   *31b*   *31c*

acetylation is due to the poor stabilization of the benzene-Ac^+ complex by an electron poor acetylbenzene moiety (*32*).

32

162

D.II Temperature Dependent Conformational Changes

In contrast to parent n°-PCP whose ^{1}H-NMR chemical shift is practically temperature indepent, lowering temperature (down to $-80\ °C$) induces a considerably large up-field shift of H_p (Fig. 9) as well as H_m absorption of 4-substituted 4°-PCP (33) [56]. This up-field shift is most significant with 33 bearing an electron withdrawing X (NO$_2$, CN, COCH$_3$) and/or a bulky X (NEt$_2$, COCH$_3$, NO$_2$). This interesting NMR feature is most probably correlated with a conformational flattening from *all face* conformer to *face^3 · lateral* conformer (Fig. 10). On the contrary the H_o absorption slightly shifts to the lower-field with the temperature decrease, in accord with the mechanism of face-lateral conformation change. Thus, the population of the laterial conformer increases at lower temperature, by

a) avoiding the steric repulsion between a bulky X and protons of other benzenes, and/or

b) stabilizing the lateral due to a dipole-π interaction (35) when X is electron withdrawing (Fig. 11).

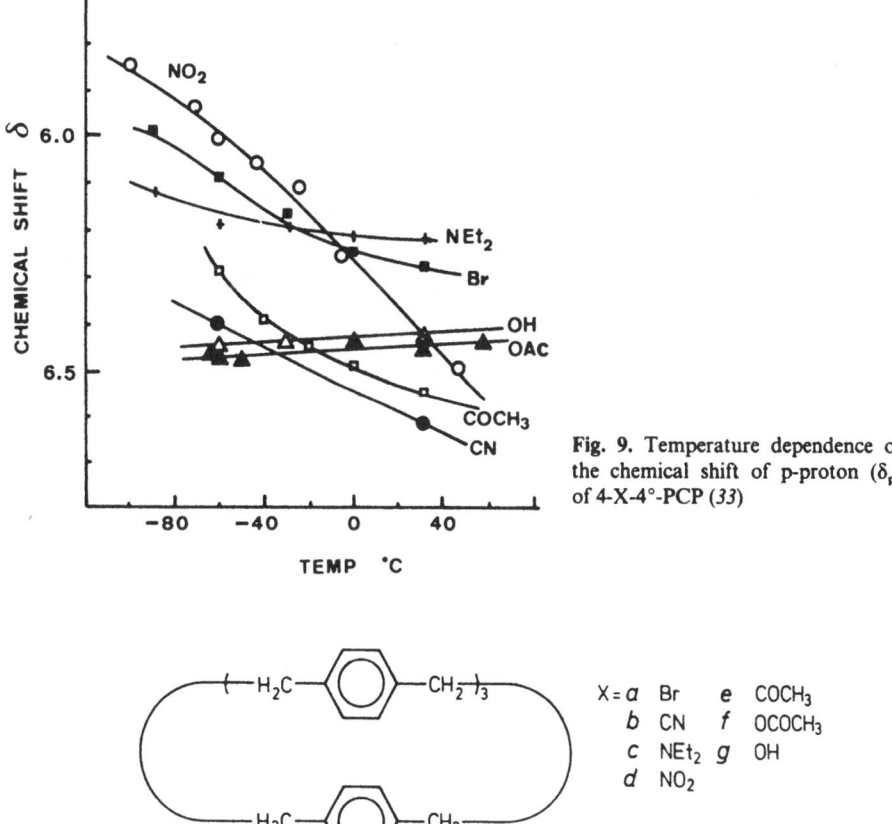

Fig. 9. Temperature dependence of the chemical shift of p-proton (δ_p) of 4-X-4°-PCP (33)

$X = a$ Br e COCH$_3$
     b CN f OCOCH$_3$
     c NEt$_2$ g OH
     d NO$_2$

33

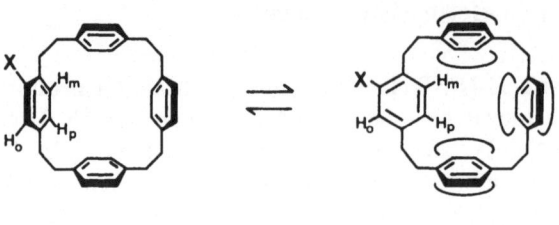

Fig. 10. Face to lateral conformation change of 4-X-4°-PCP at low temperature

34 (all face)   **35** (face³·lateral)

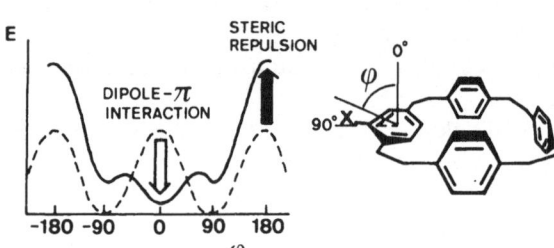

Fig. 11. Assumed potential of 4-X-4°-PCP (dotted line = 4°-PCP)

A relevant conformational flattening is found in D_2O and DMSO with Whitlock's water soluble cyclophane *18* having a substituent X = CO_2K, CO_2CH_3. From lacking up-field shift of H_6 absorption in these solvents, they proposed a skewed face (*37*, skewed F—F after Whitlock et al.) or a (face · lateral) conformer (*38*, F—E). In $CDCl_3$, however, this cyclophane preferably takes an all face conformer (*36*, or F—F) as judged by the ordinary up-field shift in a reasonable magnitude [43].

in $CDCl_3$

36   **37**   **38**

in D_2O, DMSO

$X = CO_2K$, CO_2CH_3; $Y = (CH_2)_6$

When the connecting moiety is not a flexible —$O(CH_2)_6O$— chain, but a rigid rod as dioxaoctadiyne in *17*, Whitlock et al. concluded that the (lateral²) conformer (Fig. 12a) is more consistent as the complex conformer than the (face²) conformer (Fig. 12b) with the up-field shifts in ¹H NMR. Thus, a classical "stacking" complex is much preferably formed with 2-naphthylmethyltrimethylammonium chloride than the inclusion complex [43].

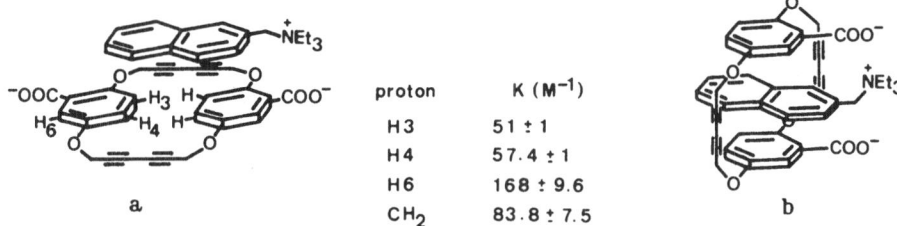

proton	K (M^{-1})
H3	51 ± 1
H4	57.4 ± 1
H6	168 ± 9.6
CH_2	83.8 ± 7.5

Fig. 12a. Stacking complex and **b** inclusion complex of a tetraoxo[8.8]paracyclophane derivative (Ref. [43])

D.III Hydrophobic Guest Binding by Cyclophanes

The basic understanding of hydrophobic binding is the destruction of the water assembly which is originally formed along the hydrophobic surface of both host and guest molecules (Fig. 1). This destruction results in a considerable amount of entropy gain (Table V).

Table V. Entropy Change for Dissolusion of Apolar Gases in Water and Some Other Solvents (cal/deg · mol, 25 °C) [65]

Gas	H_2O	CCl_4	C_2H_6
CH_4	−31.8	−14.0	−13.5
C_2H_6	−35.4	−16.8	−15.9
C_2H_4	−31.3	−16.2	−16.1
Ne	−28.8	−	−9.3
Ar	−30.2	−	−11.3

Hydrophobic guest binding in solution may be investigated spectroscopically (Table VI). A useful way to investigate hydrophobic guest binding is to use a fluorescent guest, e.g., sodium 1-anilino-8-naphthalenesulfonate (*39*, 1,8-ANS), since fluorescence intensity (very often wavelength too) is remarkably sensitive to medium polarity.

39 1,8-ANS

Table VI. Reported Methods for Investigation of Hydrophobic Interaction and Their Specifications

Method	Specification
Fluorescence	Blue shift and remarkable increase of fluorescence intensity (1,8-ANS)
Electronic absorption (UV, VIS)	π_{max} shift hypochromic shift of guest
Nuclear magnetic resonance	Shielding or deshielding due to aromatic π anisotropy (guest and/or host)

The solvent dependency of fluorescence has the following typical values:
a) fluorescence maximum

> e.g., λ_{max}^f (1,8-ANS)   472 nm in dioxane [57]
> 555 nm in water

b) fluorescence quantum yield

> e.g., φ_f (1,8-ANS)   0.65 in 99.9% dioxane-H_2O [58, 59]
> 0.008 in 20.2% dioxane-H_2O

For other hosts such as cyclodextrins [60] or proteins [57], 1,8-ANS was also successfully used.

Figure 13 is a typical example with N_4PCP (*4a*) [36]. The fluorescence maximum blue-shifted from 540 nm for the uncomplexed 1,8-ANS to 515 nm for the entirely complexed 1,8-ANS in the presence of 1×10^{-3} M N_4PCP at pH 4.01, revealing that 1,8-ANS is transfered from water into the cyclophane's cavity of less polarity quantitatively. Based on this result, the polarity of N_4PCP's cavity is approximated to that of 50% EtOH—H_2O.

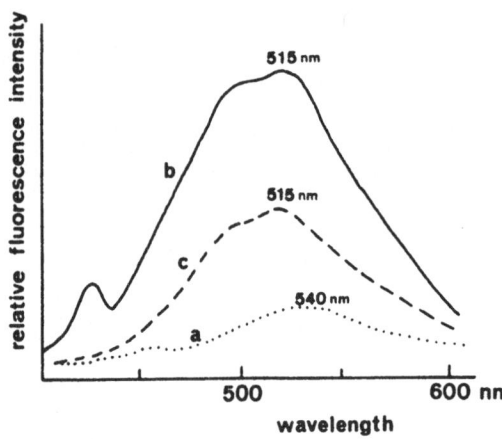

Fig. 13. Fluorescence spectrum of sodium 1-anilino-8-naphthalenesulfonate (*39*) (0.5×10^{-4} M) in (a) absence, and (b) presence of 1.0×10^{-3} M of N_4PCP (*4a*), and (c) 1.0×10^{-3} M of β-cyclodextrin (*47*) in phosphate buffer of pH 4.01/

Up to date, a considerable number of water soluble cyclophanes have been investigated on their host-guest complexes with 1,8-ANS by using fluorescence techniques (Table VII), and quantitative estimation of the host-guest association constants K is made by plotting the fluorescence increase (ΔF) against the reciprocal

Table VII. Reported Methods for Investigation of Hydrophobic Interaction between Water Soluble Cyclophanes and Guest Molecules and Reported Association Constants (K)

Host	Guest	Method	K (M^{-1})	Ref.
6°-PCP$+$CH$_2$N$^+$(CH$_3$)$_3$Cl$^-$]$_4$ (25)	Methyl orange	VIS	—	53)
N$_4$PCP · H$_3^+$	1,8-ANS (39)	Fluor.	380	36)
N$_4$PCP · H$_4^+$	1,8-ANS (39)	Fluor.	550	36)
N$_4^+$PCP (5)	1,8-ANS (39)	Fluor.	ca. 380	37)
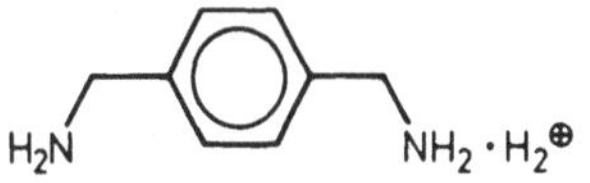	1,8-ANS (39)	Fluor.	4	37)
S$_4^+$PCP (9)	1,8-ANS (39)	Fluor.	1600	39)
S$_2^+$PCP (8)	1,8-ANS (39)	Fluor.	<50	39)
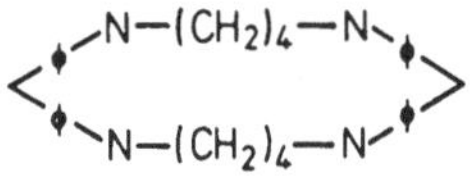 (12a)	1,8-ANS (39)	Fluor.	6250	42)
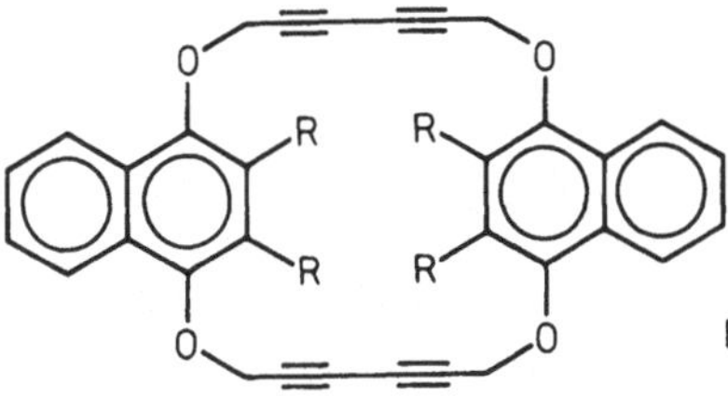	1,8-ANS	Fluor.	590	44)
R = CH$_2$OCOCH$_2$N(CH$_3$)$_2$	(20e) SO$_3$H	NMR	—	44)
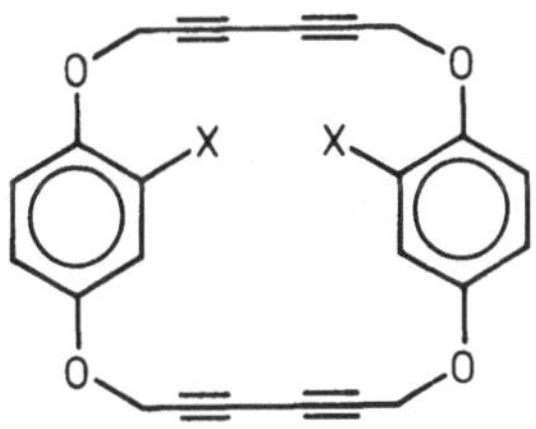 X = CO$_2$H	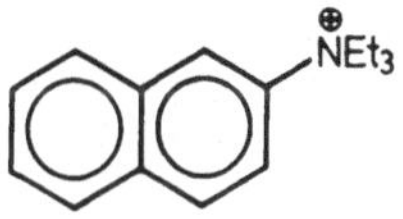 (17e) NEt$_3^{\oplus}$	NMR	51–168	43)
β-Cyclodextrin	1,8-ANS	Fluor.	57	38)
N$_4^+$PCP (5)	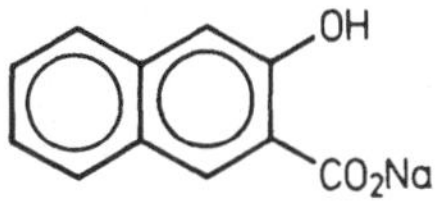 (40) OH / CO$_2$Na	UV-VIS	1130	38)
	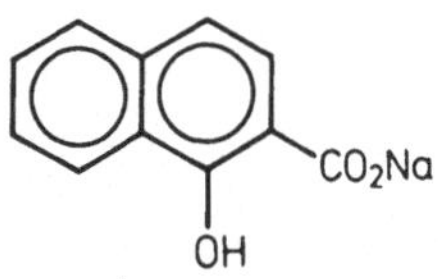 (41) CO$_2$Na / OH	UV-VIS	1150	38)

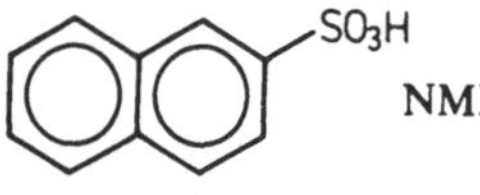

of the host concentration (C$_{host}$) by use of the following Benesi-Hildebrand relationship [61] [Eq. (16)].

$$\text{host} + \text{guest} \underset{}{\overset{K}{\rightleftharpoons}} \text{complex} \tag{15}$$

$$\frac{1}{\Delta F} = \frac{K}{\Delta f \cdot \bar{C}_{ANS} \cdot \bar{C}_{host}} + \frac{1}{\Delta f} \qquad (16)$$

As apparent from Table VII, the water soluble cyclophanes N_4PCP (*4a*), N_4^+PCP (*5*), S_4^+PCP (*9*), *12a*, or *20*, are *stronger* (better) hosts toward 1,8-ANS than β-cyclodextrin, the association constant of the latter is 57 M^{-1}.

Table VIII. Association Rate Constant (k_A) and Dissociation Rate Constant (k_D) for Substrate Binding by Several Enzymes, Water Soluble Cyclophane, or Cyclodextrin

Host	Guest		k_A (M^{-1} sec^{-1})	k_D (sec^{-1})	$K_{assoc.}$ (M^{-1})	Ref.
α-Chymo-trypsin	Furylacryloyl-L-tryptophanamide		6.2×10^6	2.7×10^3		[78]
	Proflavine		1.1×10^8	2.15×10^3		
Lysozyme	$(NAG)_2$		4.56×10^6	950		[79]
	$(NAG)_3$		4.4×10^6	28		
N_4^+PCP (*5*)[a]	*40*		1.77×10^7	1.56×10^4	1130	[38]
	41		4.40×10^7	3.8×10^4	1150	[38]
α-CD[a]	41		0.4×10^7	5.9×10^4	69	[38]
β-CD[a]	41		1.2×10^7	1.1×10^4	1130	[38]

[a] In 1/15 M phosphate buffer 0.1 M KCl, pH 7.0, 27 °C.

An enzyme binds its substrates or inhibitors very specifically and also incorporates its counterparts with extremely rapid rates — close to a diffusion controlled rate. In the water soluble cyclophane-guest complexes, we have a situation similar to the enzyme-substrate complexes. The rate process of the guest inclusion was studied, in a single case among cyclophanes, with N_4^+PCP (*5*) by using a temperature-jump technique [38], and the association (k_A) and the dissociation rate constants (k_D) are estimated (Table VIII). These rate constants reveal that the inclusion proceeds satisfactorily rapid for the ordinary hydrophobic guests. When the guest has a bulky group, both the formation and decomposition of the complex become slower. This is typically seen for cyclodextrin [62]: see Table in p. 168.

When comparing the k_A value of cyclodextrin for binding 1-hydroxy-2-naphthalene-carboxylate (*41*), the k_A value of N_4^+PCP is 1 order of magnitude larger, although the hole size of N_4PCP (5.5–7 Å) is very close to that of α-cyclodextrin (6 Å). The faster association rate for N_4^+PCP, (even faster than for β-cyclodextrin whose hole size is 7.5 Å) is reasonably ascribed to that a) somewhat more flexible nature of water soluble cyclophane (N_4^+PCP) makes it easier to have a more favorable conformation for inclusion and/or that b) positive charge of N_4^+PCP may accelerate the association [38].

Association and Dissociation Rate Constants for Inclusion Binding
by α-Cyclodextrin

Guest	k_A (s^{-1} M^{-1})	k_D (s^{-1})
R—⬡—O$^{\ominus}$	1.3×10^5	2.6×10^2
R—⬡—O$^{\ominus}$ (CH$_3$)	1.5×10^2	0.28
R—⬡—O$^{\ominus}$ (CH$_2$CH$_3$)	2.8	1×10^{-2}
R—⬡—OH (CH$_3$, CH$_2$)	no complex is formed	

$$R = {}^{\ominus}O_3S—⬡⬡—N\!=\!N—$$

These results show the very rapid attainment of a Boltzmann distribution between possible inclusion conformers required for catalysis and obtained with a certain probability, as schematically pictured in Fig. 14.

⇄ ⇄ **other binding modes**

Fig. 14. Very rapid association and dissociation in guest binding leads to a Boltzmann distribution among all possible modes of inclusion before catalysis takes place

Assuming that conformer C_i only leads to the expected catalysis among n conformers, the catalytic rate constant is given by Eq. (18).

$$p_i = \frac{[C_i]}{\sum\limits_{j=1}^{n} [C_j]} \tag{17}$$

$$k = p_i \cdot k_i = \frac{[C_i] \cdot k_i}{\sum\limits_{j=1}^{n} [C_j]} \qquad (18)$$

Therefore, a better catalyst should have a larger probability to take the i-th conformation and a larger elemental catalytic rate constant.

The absorption maximum (λ_{max}) of an ordinary dye molecule (π-π^* band) usually shifts to shorter wavelengths (very often the intensity also changes) when entirely complexed by hydrophobic cavities of cyclophanes and cyclodextrins [61-63]. For example, by adding methyl orange to 6°-PCP—[CH$_2$N$^+$(CH$_3$)$_3$]$_4$ (25) in water the wavelength was shifted from 470 nm to 450–460 nm for the complexation; in addition the intensity decreased by ca. 25% [53].

Whether a guest molecule is accomodated or not by the hydrophobic cavity between aromatic rings (face conformation) is determined by NMR (see also Fig. 12, Section D.II). The [8.8](1,4)naphthalenophane (20) hosts are in a rapid equilibrium between the syn- and anti-conformer [44]. The anti-conformer is predominant at

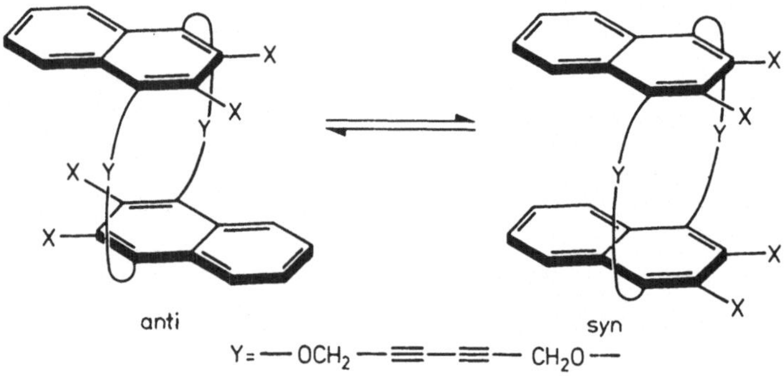

anti   syn   (20)

Y = — OCH$_2$ —≡—≡— CH$_2$O —

ordinary temperature. When 2-naphthalenesulfonic acid is used as a guest, particularly large up-field shifts were seen only for protons at 5–8 positions of both host and guest. The result was interpreted to indicate that a 2-naphthalenesulfonic acid complex (geometry as in 42) is formed by the all face-syn conformer [44].

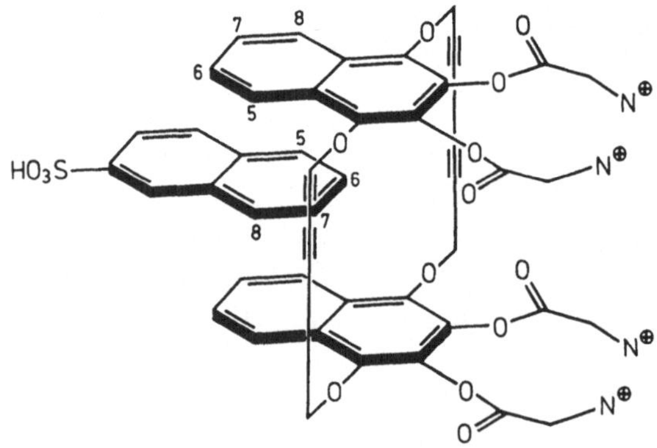

42

The Benesi-Hildebrand type treatment of Eq. (16) of the ^{1}H-NMR chemical shift change was used to determine the association constant $K = 51–168 \ M^{-1}$ for the stacking complex by hexadiynyl ether linked paracyclophane *17e* (Table VII, Fig. 12).

D.IV Crystalline Complex Formation

Table IX. Crystalline Complexes/Clathrates Formed by Cyclophanes

Host	Guest	m	n	Ref.
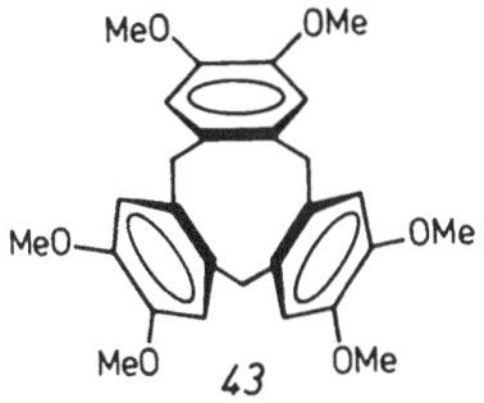 *43*	CH_3CH_2OH $CH_3COCH_2CH_3$ $CH_3COOCH_2CH_3$	1 1 1	1.5 3.2 1.6	27)
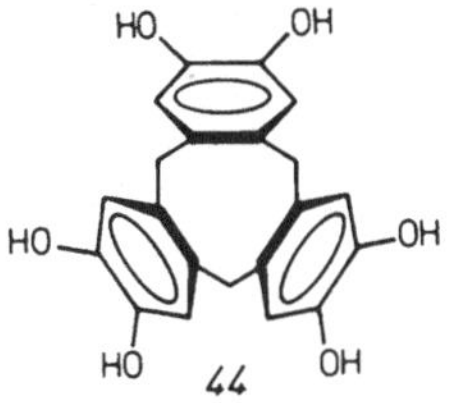 *44*	CH_3COCH_3 $CH_3CH(OH)CH_3$ CH_3SOCH_3 $HCON(CH_3)_2$ $CH_3CON(CH_3)_2$ etc.	1 1 1 1 1	2.0 2.0 3.0 3.1 3.0	27)
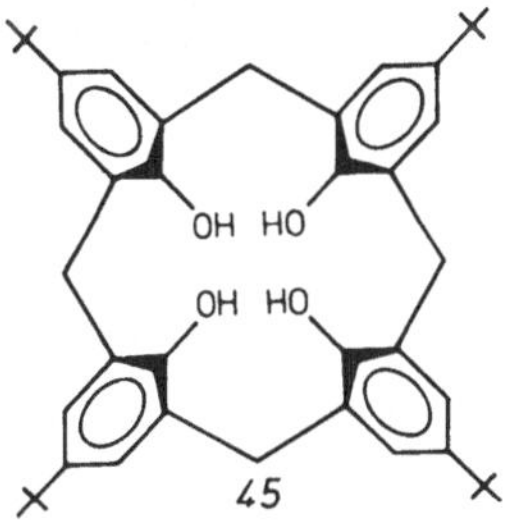 *45*	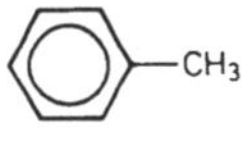	1	1	31)
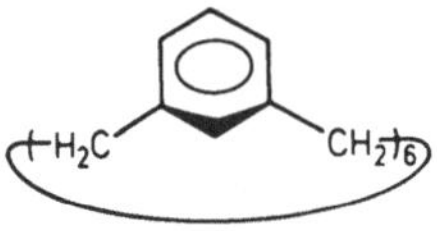 *46*	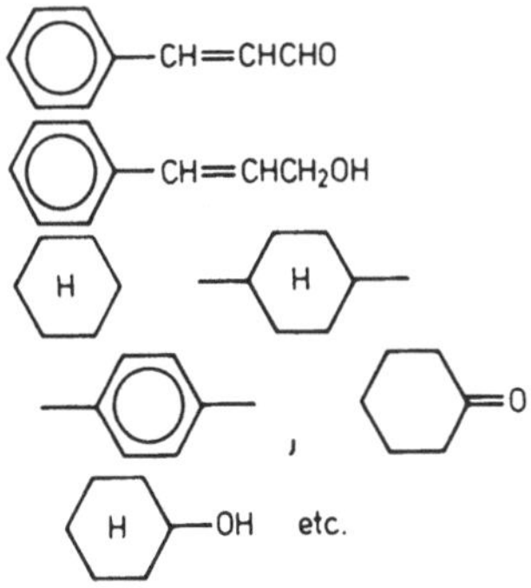			32,33)

* No X-ray study.

A very limited number of host-guest pairs has been studied regarding the crystalline structure, and most of the crystalline complexes so far known are "lattice (cavity) inclusion", $(Host)_m \cdot (Guest)_n$ (type b, Table XI) in which the ratio m:n generally varies with the sort of guest molecule.

Crystal Lattice Inclusion $(Host)_m \cdot (Guest)_n$

It is long known that a so-called clathrate compound is formed from urea, thiourea, or perhydrotriphenylene etc. with a variety of guest compounds. Typical hosts forming such clathrate compounds are cyclotriveratrylene (*43*) and cyclotricatechylene (*44*), and their guest-selectivity is relatively poor (Table IX). More recently, calix[4]arene *45* has been found by X-ray to give a (1:1) clathrate with toluene [31]. The Practically no or little guest-selectivity found with 6°-metacyclophane *46* is most likely due to the somewhat more flexible nature of *46*, although no X-ray study is reported.

In lattice inclusion, host-host interaction (very often host-guest interaction too) should be significant, affording a cage or a channel for the guest accomodation. This was demonstrated by X-ray with cyclotricatechylene-2-propanol clathrates, in which hydrogen bondings between "cone" like hosts exist in addition to host-guest hydrogen bondings (Fig. 15).

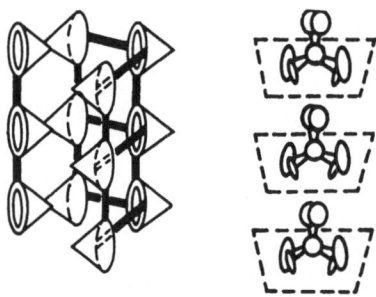

Fig. 15. Schematic representation of lattice (cavity) inclusion of cyclotricatechylene-2-propanol clathrate (thick line represents a hydrogen bonding)

Fig. 16. Molecular cavity inclusion of $N_4PCP \cdot CHCl_3$

Molecular Cavity Inclusion $(Host \cdot Guest)_m$

The formation of a crystalline complex by our N_4PCP (*4a*) is rather specific to a limited number of guest molecules such as $CHCl_3$ or CH_2Cl_2; $-\Delta G$ (Host-Guest) seems significantly important (via nearly maximal van der Waals contact, Fig. 3a). With poorly fitting guest molecules — such as 2-methylbutanol — molecular inclusion was not observed at all but only homogeneous host crystals $(Host)_m$ were obtained [54]. Thus, by maximal van der Waals interaction we have 1:1 molecular inclusion crystal (Table X and Fig. 16).

Therefore, it seems very appropriate to classify crystalline complexes from macrocyclophanes into the three categories shown in Table XI.

Table X. One to One Molecular Cavity Inclusion Complexes of Macrocyclophanes

Host	Guest	Space group	Ref.
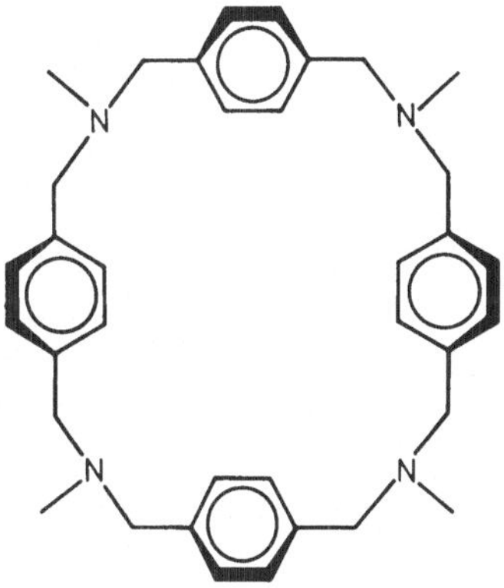 CHCl$_3$	CHCl$_3$	C2 (chiral)	[54]
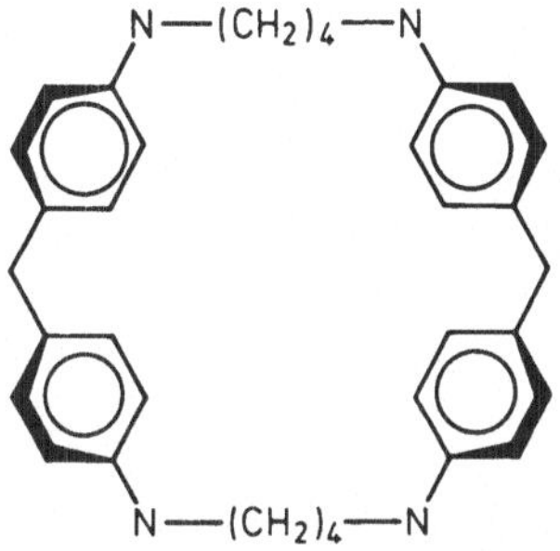		p2$_1$/n	[42]

Table XI. Classification of Crystalline Complexes Formed from Macrocyclophane Host with Guest

Solution	General formula	Type
Host$_{solv}$ + Guest$_{solv}$	a) (Host · Guest)	molecular cavity inclusion
	b) (Host)$_m$ · (Guest)$_n$	lattice cavity inclusion
	c) (Host)$_m$ + Guest$_{solv}$	homogeneous crystal

E Hydrophobic Complex Formation

E.I Hydrophobic Guest Binding by Cyclodextrins

Cyclodextrins are natural cyclic oligomers of D-glucose (α-CD, n = 6; β-CD, n = 7; γ-CD, n = 8; etc.). They are prepared by specific enzymes; their shape is schematically depicted in *47*, as a cyclindrical torus. The interior is apolar (due to many C—H, C—C and C—O bonds) and the cavity size (ca. 6 Å for α-CD, ca. 7.5 Å for β-CD) is large enough to accomodate benzenes or naphthalenes. Thus, cyclodextrin is a class of naturally occurring *water soluble hosts*, whose host-guest feature has been well documented [2-6]. For this reason, discussions on hydrophobic interactions as the principal driving force for the guest binding by CD should only be briefly mentioned here.

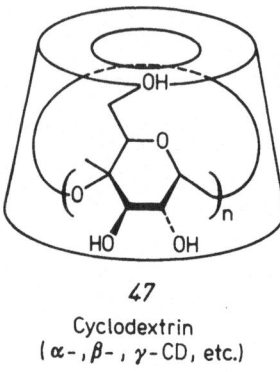

47

Cyclodextrin
(α-, β-, γ-CD, etc.)

Certain number of water molecules occupy the cavity of crystalline cyclodextrin, e.g., 2 H$_2$O in α-CD or 9 H$_2$O in β-CD (from X-ray) [66,67]. However, a specific guest molecule, when added into cyclodextrin solution, drives these water molecules out of the cavity, as demonstrated by X-ray with inclusion complexes or substituted cyclodextrins, e.g., mono-tert-butylsulfenyl-β-cyclodextrin (Fig. 17) [68].

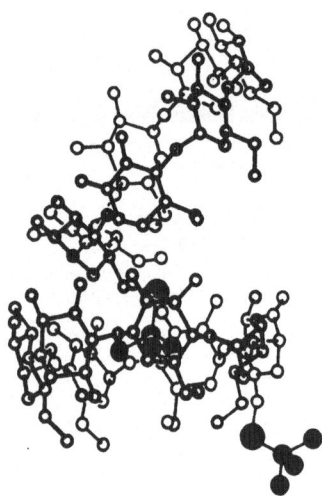

Fig. 17. X-Ray crystal structure of mono-*tert*-butylsulfenyl-β-cyclodextrin

Flexibly capped cyclodextrins [69] or rigidly capped cyclodextrins [60,70] have been prepared so that a hydrophobic moiety, appended or capped onto the primary rim of β-CD, extends the hydrophobic surface area of cyclodextrin (Fig. 18b). They remarkably enhanced the association with most hydrophobic guests; e.g., the association constants (M^{-1}) for 1,8-ANS with β-CD is *58*, with terephthalate cap *640*, and with diphenylmethanedisulfonate cap *1300* [60].

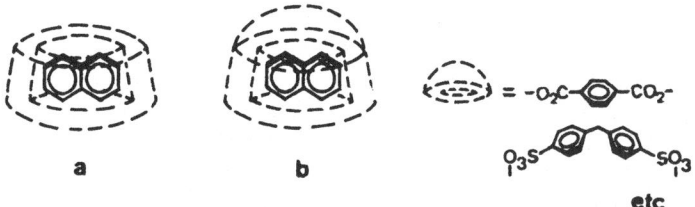

a   **b** $= -O_2C-\langle\bigcirc\rangle-CO_2^-$

$O_3S-\langle\bigcirc\rangle\langle\bigcirc\rangle-SO_3^-$

etc

Fig. 18. Hydrophobic capping (**b**) increases the surface area of the hydrophobic interior of cyclodextrin, enhancing the contact with a hydrophobic guest molecule

E.II Thermodynamics of Hydrophobic Guest Binding

Accoding to the concept of hydrophobic interaction, we have calculated the total interaction energy (ΔG°) between α-cyclodextrin and some benzene derivatives based on the following thermodynamic process [71a] [Eqn. (19–20), see Fig. 19]:

$$\Delta H_{inclusion} = (H^c_{vdW} - H^w_{vdW}) + (H^c_{conf} - H^w_{conf})$$
$$- \Delta H^g_{cluster} - 2\Delta H^w_{vap} - 2H_{H\text{-bond}} \tag{19}$$

$$\Delta S_{inclusion} = (S^g_{rot(1-D)} - S^g_{rot(3-D)} - S^g_{trans})$$
$$- 2(S^w_{rot(3-D)} - S^w_{trans}) + 2\Delta S^w_{gas \to liq} - \Delta S^g_{cluster} \tag{20}$$

This thermodynamic picture (Fig. 19) seems most appropriate for understanding *hydrophobic inclusion* as apparent from the calculated driving force energy [−ΔG° based on Eq. (19)–(20)] in good accord with the observed values (Table XII) [71].

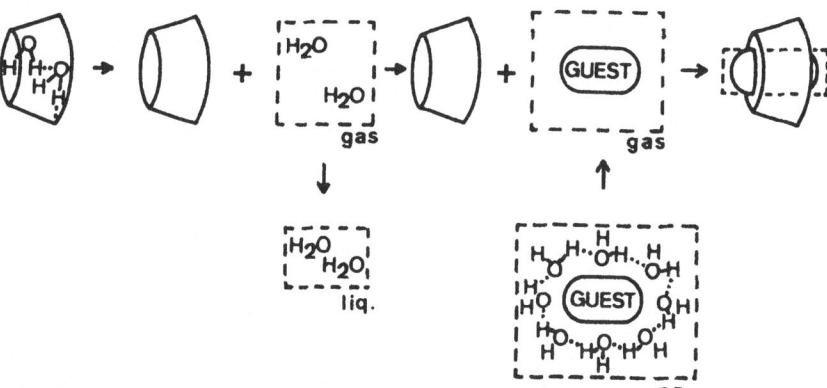

Fig. 19. Hypothetical thermodynamic presentation of inclusion complex formation by cyclodextrin

The conformation change during the inclusion process is not too serious (Table XIII, XIV). Any macroring strain should not be very important, since no drastic drop of p-nitrophenol binding was observed by Bergeron et al. with hexakis(2,6-di-O-methyl)-α-CD, in which hydrogen bondings are prevented to distort the ring or to let glucopyranose rings take orthogonal positions with hydrogen bonding [76].

Table XII. Calculated Enthalpy Change, Entropy Change, and Free-Energy Change in Inclusion Complex Formation by α-Cyclodextrin

Guest	$\Delta H_{inclusion}$	$-T\,\Delta S_{inclusion}$	$\Delta G_{inclusion}$	
			Calcd	Obsd
Benzene	−3.99	−0.51	−4.50	
p-Iodoaniline	−7.35	−1.64	−8.99	−5.9
Methyl orange	−6.53	+0.33	−6.20	−5.1

ᵃ $Kcal \cdot mol^{-1}$, 25 °C.

Table XIII. Calculated Values of Enthalpy Change (kcal/mol) in Inclusion Complex Formation

Guest	$H_{vdW}^c - H_{vdW}^w$	$H_{conf}^c - H_{conf}^w$	$-\Delta H_{cluster}^g$	$-2\,\Delta H_{vap}^w$	$-2H_{H\text{-}bond}$
Benzene	−3.75	+4.38	+4.10	−20.92	+12.2
p-Iodoaniline	−10.09	+4.38	+7.08	−20.92	+12.2
Methyl orange	−8.35	+4.38	+6.71	−20.92	+12.2

Table XIV. Calculated Values of Entropy Changes (cal/deg · mol) in Inclusion Complex Formation

Guest	$S_{rot(1-D)}^g - S_{rot(3-D)}^g$		$-2(-S_{rot(3-D)}^w - S_{trans}^w + 2\,\Delta S_{gas \to liq}^g$
	$-S_{trans}^g$	$-\Delta S_{clust}^g$	
Benzene	−52.0	+20.1	+33.6
p-Iodoaniline	−62.9	+34.8	+33.6
Methyl orange	−67.6	+32.9	+33.6

By contrast, thermodynamic terms, ($\Delta H_{cluster}$, ΔH_{vap}^w, $\Delta S_{cluster}^g$, ΔS^w, $\Delta S_{gas \to liq}^w$) coming from the change of the state of water molecules in the cavity and around the apolar guest, are much more important compared with other contributions (Table XIII, XIV). It seems reasonable to assume that the thermodynamic approach with Eq. (19)–(20) is equivocally applicable to hydrophobic guest binding in general, e.g., protein, cyclophane etc.

E.III Crystalline Water Network around Apolar Molecules

It is still an open question what is the nature of the water assembly around any apolar surface in water. Unfortunately, experimental informations are very limited [72]. Current theoretical studies — statistical thermodynamics, Monte Carlo or Molecular Dynamics simulations — reveal that such a water cluster has a rather small size and a very short lifetime [73, 74].

For evaluating hydrophobic interaction energy experimentally, we have counted the number of water molecules around apolar molecules [75a]. For purpose, the hydrophobic hydration (ΔH^w, ΔS^w) is evaluated in a way consistent to experimental ΔH_{diss} and ΔS_{diss} values for the dissolution of an apolar gas in water [Eqs. (21), (22)], giving the number of water molecules n (Table XV). It is very interesting

$$\Delta H_{diss} = n(\Delta H^w + \Delta H^{vdW}) \tag{21}$$

$$\Delta S_{diss} = n \, \Delta S^w \tag{22}$$

that the water assembly around each apolar gas should have a rather small size to reconcile experimental data, and the van der Waals energy seems to be also significant in addition to hydrogen bondings for hydrophobic hydration.

Table XV. Number of Water Molecules around an Apolar Gas

Guest	n (25 °C)
He	7.0
Ne	7.6
Ar	8.9
Kr	9.3
H_2	7.3
N_2	8.9
O_2	8.9

The number of water molecules in the *crystalline* water clathrate of an apolar gas was also counted recently by adopting an interesting approach of computer construction of water network [75].

F Catalysis via Substrate Complex

F.I Substrate Specificity in Hydrolysis Rate Acceleration

If a catalytic site has a strong polar group (e.g. $^+NR_4$ as in N_4^+PCP), recognition becomes precise and strict, just like enzyme recognition. Detailed investigation of the hydrolysis activity of N_4^+PCP (5) for aryl esters demonstrated that the catalyst recognizes even a small structural change as shown in Fig. 20. In fact, two important and interesting enzyme-like characteristics are found:

a) Michaelis-Menten mechanism [Eq. (23)],

$$\text{aryl-O-}\overset{\overset{\text{O}}{\|}}{\text{C}}\text{-CH}_2\text{Cl} + N_4^+\dot{P}CP \underset{k_{-1}}{\overset{k_1}{\rightleftharpoons}} \text{CS complex} \xrightarrow{k_{cat}} N_4^+\dot{P}CP \quad (23)$$

$$48a-c \qquad\qquad\qquad\qquad\qquad\qquad\qquad + \text{aryl-OH}$$

$$\Big\downarrow k_0 \qquad\qquad\qquad\qquad\qquad\qquad\qquad\qquad + \text{HOCOCH}_2\text{Cl}$$

$$\qquad\qquad\qquad K_m = (k_{-1} + k_{cat})/k_1$$

$$\text{products}$$

b) selective acceleration to the guest shape (β-naphthyl selectivity) (Table XVI) [38].

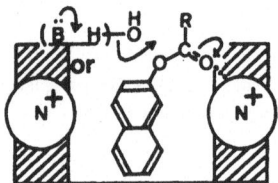

Fig. 20. Stabilization of the anionic transition state by N_4^+PCP (5) in the catalyzed ester hydrolysis

An inclusion conformer leading to a certain catalysis (Fig. 14) is allowed by a certain probability (Section D-III), and the magnitude of the association ($1/K_m^{c_i}$ or more appropriately, K_{assoc}) of the i-th conformer is correlated with the apparent (observed) association ($1/K_m^{app}$, K_{assoc}^{app}) as in Eq. 24, where f_{c_i} is the fraction of the i-th conformer.

$$\frac{1}{K_m^{app}} = \sum_i \left(\frac{1}{K_m^{c_i}} \cdot f_{c_i} \right) \qquad (24)$$

$$k_{obs} = \left(\frac{1}{K_m^{app}} \right) \cdot f^{npb} \cdot k^{npb} + \left(\frac{1}{K_m^{app}} \right) \cdot f^{pb} \cdot k_{cat}^{pb} . \qquad (25)$$

Therefore, the observed rate constant is given by Eq. (25), and the excellent inclusion catalysis by N_4^+PCP (5) is understood by a larger fraction of productive binding (f^{pb}) than the non-productive f^{npb}, especially toward β-naphthyl ester, a specific substrate.

For CTAB micellar catalysis, however, f^{pb} and/or k_{cat}^{pb} should be very small, since CTAB micelles have 1/3.6 times smaller k_{cat} value (Table XVII) than N_4^+PCP (5), whereby no substrate specificity was observed.

Table XVI. Shape Sensitive Rate Acceleration (k_{cat}/k_0) in Hydrolysis of Chloroacetates Catalyzed by N_4^+PCP (5)

	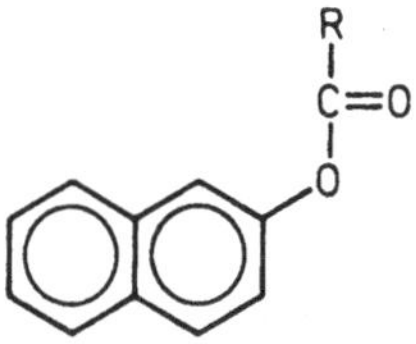 48c	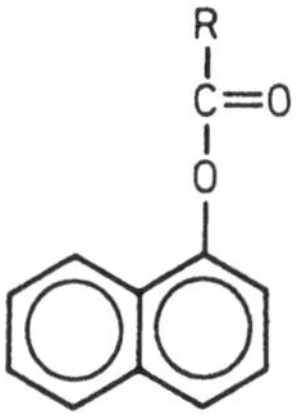 48b	 48a (NO$_2$)	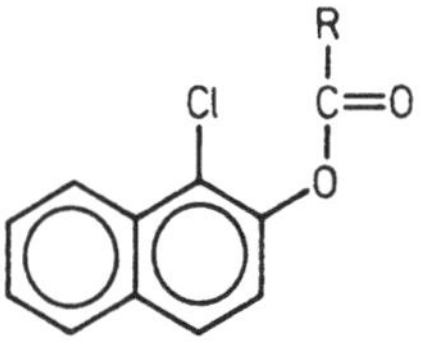 49
$\dfrac{k_{cat}}{k_0}$	25	6.0	2.6	0.085

Table XVII. Catalytic Parameters in Ester Hydrolysis

Catalyst	Substrate	pH	k_{cat} (10^{-3} s^{-1})	K_m(mM)
5	48c	8.10	19.2	0.54
	48b		4.9	0.18
	48a		14.6	0.51
50	48a	6.65	90	2.3
	51 (O$_2$N—C$_6$H$_3$(NO$_2$)—OCOCH$_2$Cl)		1830	1.5
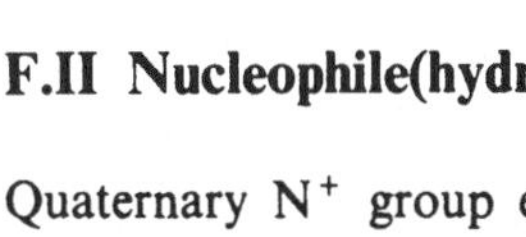	48c	8.10	1.3	2.6
CTAB	48c	8.25	5.9	0.03
	48b		4.5	0.02

F.II Nucleophile(hydroxamate)-Inclusion-Electrostatic Catalysis

Quaternary N^+ group embedded in cyclophane skeleton was effective and more selective than CTAB micellar catalysis for guest hydrolysis (Fig. 20). To gain further acceleration, a strong nucleophile, hydroxamate functional group was introduced [Eq. (26)]. In this rate enhancement, the transition state stabilization seems to be so significant (formation of tetrahedral intermediate is so fast) that the catalysis becomes markedly sensitive to the leaving ability of aryloxyl anion (see Broensted $\beta_{LG} = -0.6$ for phenol of pK$_a$ > 7, Fig. 21).

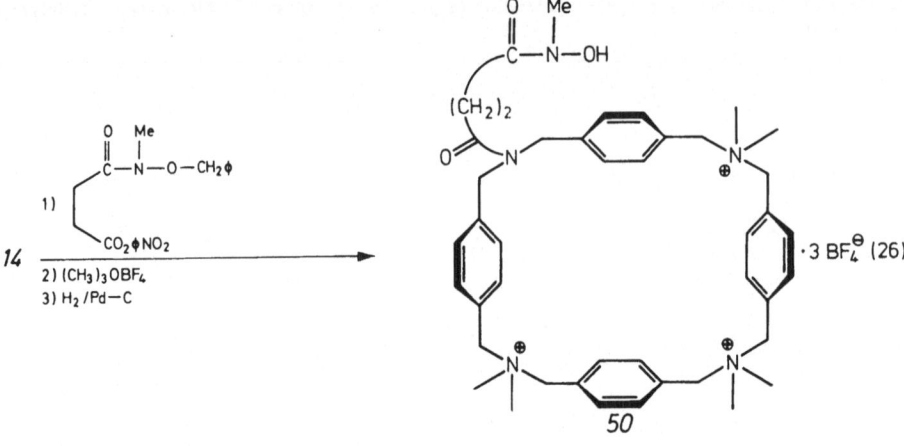

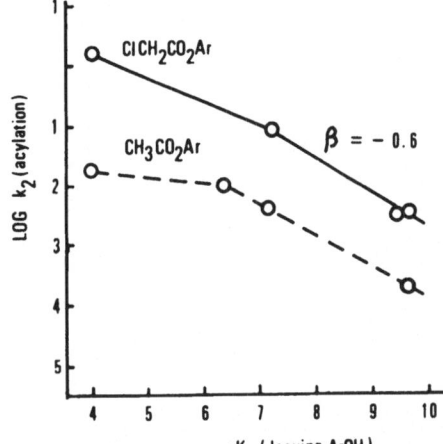

Fig. 21. Broensted-type plot of nucleophilic-electrostatic inclusion catalysis by *50* in ester hydrolysis

A reasonable picture is the appropriate recognition of the substrate by nucleophile-(hydroxamate) as well as electrostatic(stabilization) recognition site as schematically shown in *51*.

G Concluding Remarks and Outlook

The inclusion of substrates into synthetic cavities is generally acknowledged today. We can utilize several interactions — hydrophobic interaction, charge-charge and/or charge-dipole interaction — for the binding design. In doing so, cyclophane systems are of particular interest. Such compounds bind hydrophobic guests with significant conformational changes — induced by the host-guest interaction —, which sometimes affect the host-host interaction as well in the crystalline state. A macrocyclophane host of conformational flexible nature now offers the opportunity to model "induced fit". Their investigation might be one of the promising ways to achieve much stronger and more specific guest recognition.

Molecular cavity inclusion in the crystalline phase, which is one of the newest aspects in the host-guest chemistry of macrocyclophanes, promises future development especially for detailed understanding of weak intermolecular interaction.

H References

1. Cram, D. J., Cram, J. M.: Science *183*, 803 (1974)
2. Cramer, F.: Einschlußverbindungen, Berlin: Springer (1954)
3. Saenger, W.: Environment Effect on Molecular Structure and Properties (Pullman, B. ed.), Dordrecht: Reidel (1976)
4. Breslow, R.: Studies on Enzyme Models, Advances in Chemistry Series No. 100, Amer. Chem. Soc., Washington, D.C. (1971)
5. Bender, M. L., Komiyama, M.: Cyclodextrin Chemistry, Berlin: Springer (1978)
6. Tabushi, I.: Accounts Chem. Res. *15*, 66 (1982)
7. Cram, D. J.: Application of Biochemical Systems in Organic Chemistry, (Jones, J. B. ed.), New York: Wiley (1976)
8. Lehn, J. M.: Accounts Chem. Res. *11*, 49 (1978)
9. Stetter, H., Roos, E. E.: Chem. Ber. *88*, 1390 (1955)
10. Stetter, H., Marx-Moll, L.: ibid. *91*, 677 (1958)
11. Stetter, H., Rutzen, H.: ibid. *91*, 1775 (1958)
12. Stetter, H., Mayer, K. H.: ibid. *94*, 1555 (1961)
13. Stetter, H., Mayer, K. H.: ibid. *91*, 1982 (1958)
14. Stetter, H., Wirth, W.: Liebigs Ann. Chem. *631*, 144 (1960)
15. Stetter, H., Roos, E. E.: Chem. Ber. *87*, 566 (1954)
16. Smith, B. H.: Bridged Aromatic Compounds, New York: Academic Press (1964)
17. Christol, H. et al.: Tetrahedron Lett., 2591 (1979)
18. Bottino, F. et al.: ibid. 1171 (1979)
19. Danieli, R., Ricci, A., Ridd, J. H.: J. Chem. Soc. Perkin II 3, 290 (1976)
20. New Kome, G., Nayak, A.: J. Org. Chem. *43*, 409 (1978)
21. Tam, T-F. et al.: ibid. *41*, 1289 (1976)
22. Sato, T., Uno, K.: J. Chem. Soc. Perkin I 9, 895 (1973)
23. Kyba, E. P. et al.: J. Amer. Chem. Soc. *102*, 139 (1980)
24. Vögtle, F., Brombach, D.: Chem. Ber. *108*, 1682 (1975)
25. Vögtle, F., Lichtenthaler, R. G.: Angew. Chem. *84*, 588 (1972)
26. Vögtle, F., Müller, W. M.: Angew. Chem. *94*, 138 (1982) Intern. Edit. *21*, 147 (1982)
27. Hyatt, J. A. et al.: J. Org. Chem. *45*, 5074 (1980)
28. Lindsey, A. S.: J. Chem. Soc. 1685 (1965)
29. Goldup, A., Morrison, A. B., Smith, G. W.: ibid. 3864 (1965)
30. Gaglioti, V. et al.: J. Inorg. Nucl. Chem. *8*, 572 (1958)
31. Andreetti, G. D., Ungaro, R., Pochinic, A.: J. Chem. Soc., Chem. Commun., 1005 (1979)
32. Ichikawa, Y. et al.: Chem. Abstr. *87*, P 134101 p (1977); *89*, P 197137 s (1978)
33. Yamaji, T., Yoshisato, E., Hiramatsu, T.: Chem. Abstr., *89*, P 42589 s (1978); *89*, P 6017 v (1978); *89*, P 42730 f (1978)
34. Nemethy, G., Scheraga, H. A.: J. Chem. Phys. *36*, 3401 (1962)
35. Urushigawa, Y., Inadzu, T., Yoshino, T.: Bull. Chem. Soc. Japan *44*, 2546 (1971)

36. Tabushi, I., Kuroda, Y., Kimura, Y.: Tetrahedron Lett., 8827 (1976)
37. Tabushi, I., Kimura, Y., Yamamura, K.: J. Amer. Chem. Soc. *100*, 1304 (1978)
38. Tabushi, I., Kimura, Y., Yamamura, K.: ibid. *103*, 6486 (1981)
39. Tabushi, I., Sasaki, H., Kuroda, Y.: ibid. *98*, 5727 (1976)
40. Imashiro, F. et al.: Tetrahedron Lett., 371 (1976)
41. Tabushi, I., Kimura, Y., Yamamura, K.: Chemical Approaches to Understanding Enzyme Catalysis, (Green, B. S., Ashani, Y., Chipman, D. eds.), Amsterdam: Elsevier (1982)
42. Odashima, K. et al.: J. Amer. Chem. Soc. *102*, 2504 (1980)
43. Jarvi, E. T., Whitlock, Jr., H. W.: ibid. *102*, 657 (1980)
44. Adams, S. P., Whitlock, H. W.: *104*, 1602 (1982)
45. Baker, W., McOmie, J. F., Norman, J. M.: J. Chem. Soc., 1114 (1951)
46. Tabushi, I. et al.: Tetrahedron *27*, 4845 (1971)
47. Tabushi, I. et al.: ibid. *28*, 3381 (1972)
48. Tabushi, I. et al.: J. Org. Chem. *40*, 1946 (1975)
49. Müller, E., Röscheisen, G.: Chem. Ber. *90*, 543 (1957)
50. Schaefer, J. P.: J. Polymer Sci. *15*, 203 (1955)
51. Errede, L. A., Cassidy, J. P.: J. Amer. Chem. Soc. *82*, 3653 (1960)
52. Tabushi, et al.: Catalyst (Japanese) *14*, 147 (1972)
53. Tabushi, I., Kuroda, Y.: ibid. *16*, 78 (1974)
54. Tabushi, I. et al.: Organic Structural Chemistry Symp., Kyoto (1982)
55. Imashiro, F., Yoshida, Z., Tabushi, I.: Tetrahedron *29*, 3521 (1973)
56. Tabushi, I., Yamada, H.: ibid. *33*, 1101 (1977)
57. Turner, D. C., Brand, L.: Biochem. *7*, 338 (1968)
58. Kosower, E. M., Dodiak, H.: J. Amer. Chem. Soc. *100*, 4173 (1978)
59. Kosower, E. M., Dodiak, H., Kanety, H.: ibid. *100*, 4179 (1978)
60. Tabushi, I. et al.: ibid. *98*, 7855 (1976)
61. Cramer, F.: Chem. Ber. *84*, 851 (1951)
62. Cramer, F., Saenger, W., Spatz, H. C.: J. Amer. Chem. Soc. *89*, 14 (1967)
63. Tabushi, I. et al.: ibid. *101*, 1019 (1979)
64. (a) Kaiser, E. T., Kaiser, B. L.: Accounts Chem. Res. *5*, 219 (1972)
    (b) Halford, S. E.: Biochem. J. *149*, 411 (1975)
    (c) Taylor, P. W., King, R. W., Burger, A. S. V.: Biochem. *9*, 2638 (1970)
65. Frank, H. S., Evans, M. W.: J. Chem. Phys. *74*, 170 (1970)
66. Manor, P. C., Saenger, W.: Nature *237*, 392 (1972)
67. Linder, K., Saenger, W.: Angew. Chem., Int. Ed. Engl. *9*, 17 (1978)
68. Hirotsu, K. et al.: J. Org. Chem. *47*, 1143 (1982)
69. Emert, J., Breslow, R.: J. Amer. Chem. Soc. *97*, 670 (1975)
70. (a) Beslow, R. et al.: ibid. *100*, 3227 (1978)
    (b) Breslow, R., Bovy, P., Hersh, C. L.: ibid. *102*, 2115 (1980)
    (c) Tabushi, I. et al.: ibid. *103*, 2017 (1982)
    (d) Tabushi, I. et al.: ibid. *104*, 2017 (1982)
71. Tabushi, I. et al.: ibid. *100*, 916 (1978)
72. Eisenberg, D., Kauzmann, W.: The Structure and Properties of Water, Oxford: Clarendon (1969)
73. Hagler, A. T., Scheraga, H. A., Nemethy, G.: Ann. New York Acad. Sci., 51 (1973)
74. (a) e.g. Baker, J. A., Watts, R. O.: Chem. Phys. Lett. *3*, 144 (1969)
    (b) e.g. Rahman, A., Stillinger, F. H.: J. Chem. Phys. *55*, 3336 (1971)
75. Tabushi, I., Kiyosuke, Y., Yamamura, K.: Bull. Chem. Soc. Japan *54*, 2260 (1981)
76. Bergeron, R. J., Meeley, M. P.: Bioorganic Chem. *5*, 197 (1976)
77. In the meantime when preparing this manuscript, an X-ray study on N_4PCP · dioxane appeared by Abbott, S. J. et al.: J. Chem. Soc., Chem. Commun., 796 (1982)
78. (a) Havsteen, B. H.: J. Biol. Chem. *242*, 769 (1967)
    (b) Hess, G. P. et al.: Phil. Trans. Roy. Soc. *B256*, 27 (1969)
    (c) Sykes, B. D.: J. Amer. Chem. Soc. *91*, 949 (1969)
79. (a) Chipman, D. M., Schimmel, P. R.: J. Biol. Chem. *243*, 3771 (1968)
    (b) Sykes, B. D.: Biochem. *8*, 1110 (1969)
80. Binsch, G.: Topics in Stereochemistry, *3*, 97 (1970)

Author Index Volumes 101–113

Contents of Vols. 50–100 see Vol. 100
Author and Subject Index Vols. 26–50 see Vol. 50

The volume numbers are printed in italics

Venugopalan, M., and Vepřek, S.: Kinetics and Catalysis in Plasma Chemistry. *107*, 1–58 (1982).
Vepřek, S., see Venugopalan, M.: *107*, 1–58 (1983).
Vögtle, F., see Rossa, L.: *113*, 1–86 (1983).
Vostrowsky, O., see Bestmann, H. J.: *109*, 85–163 (1983).
Voronkov. M. G., and Lavrent'yev, V. I.: Polyhedral Oligosilsequioxanes and Their Homo Derivatives. *102*, 199–236 (1982).

Wachter, R., see Barthel, J.: *111*, 33–144 (1983).
Wilke, J., see Krebs, S.: *109*, 189–233 (1983).

Yamamura, K., see Tabushi, I.: *113*, 145–182 (1983).

Zollinger, H., see Szele, I.: *112*, 1–66 (1983).

Applied Physics A Solids and Surfaces

Applied Physics A "Solids and Surfaces" is devoted to concise accounts of experimental and theoretical investigations·that contribute new knowledge or understanding of phenomena, principles or methods of applied research.

Emphasis is placed on the following fields:

Solid-State Physics
Semiconductor Physics: **H.J.Queisser**, MPI Stuttgart
Amorphous Semiconductors: **M.H.Brodsky**, IBM Yorktown Heights
Magnetism and Superconductivity: **M.B.Maple**, USCD, La Jolla
Metals and Alloys, Solid-State Electron Microscopy: **S.Amelinckx**, Mol
Positron Annihilation: **P.Hautojärvi**, Espoo
Solid-State Ionics: **W.Weppner**, MPI Stuttgart

Surface Science
Surface Analysis: **H.Ibach**, KFA Jülich
Surface Physics: **D.Mills**, UC, Irvine
Chemisorption: **R.Gomer**, U. Chicago

Surface Engineering
Ion Implantation and Sputtering: **H.H.Andersen**, U. Copenhagen
Laser Annealing and Processing: **R.Osgood**, Columbia U.
Integrated Optics, Fiber Optics, Acoustic Surface Waves: **R.Ulrich**, TU Hamburg
Device Physics: **M.Kikuchi**, Sony Yokohama

Coordinating Editor: **H.K.V.Lotsch**, Heidelberg

Special Features:
– Rapid publication (3–4 months)
– No page charges for concise reports
– 50 complimentary offprints

Subscription information and/or **sample copies** are available from your bookseller or directly from Springer-Verlag, Journal Promotion Dept., P.O.Box 105280, D-6900 Heidelberg, FRG

Springer-Verlag
Berlin
Heidelberg
New York
Tokyo

V. N. Kondratiev, E. E. Nikitin

Gas-Phase Reactions

Kinetics and Mechanisms
1981. 1 portrait, 64 figures, 15 tables. XIV, 241 pages
ISBN 3-540-09956-5

Contents: General Kinetic Rules for Chemical Reactions. – Mechanisms of Chemical Reactions. – Theory of Elementary Processes. – Energy Exchange in Molecular Collisions. – Unimolecular Reactions. – Combination Reactions. – Bimolecular Exchange Reactions. – Photochemical Reactions. – Chemical Reactions in Electric Discharge. – Radiation Chemical Reactions. – Chain Reactions. – Combustion Processes. – References. – Subject Index.

G. Kortüm

Reflectance Spectroscopy

Principles, Methods, Applications
Translator from the German: J. E. Lohr
1969. 160 figures. VI, 366 pages
ISBN 3-540-04587-2

G. V. Vinogradov, A. Y. Malkin

Rheology of Polymers

Viscoelasticity and Flow of Polymers
1980. 220 figures, 3 tables. XII, 467 pages
Moscow: Mir Publishers. ISBN 3-540-09778-3

Contents: Basic Concepts of Rheology. – Shear Viscosity. – Viscoelastic Properties of Polymer Melts and Solutions. – Normal Stresses in Shear (the Weissenberg Effect). – The Rubber-Like Behaviour of Polymers in Flow. – Rheological Properties of Filled Polymers. – Uniaxial Extension. – Author Index. – Subject Index.

M. Mehring

Principles of High Resolution NMR in Solids

2nd revised and enlarged edition. 1983. VIII, 342 pages
ISBN 3-540-11852-7
(Originally published in the series NMR – Basic Principles and Progress, Vol. 11)

Contents: Introduction. – Nuclear Spin Interactions in Solids. – Multiple-Pulse NMR Experiments. – Double Resonance Experiments. – Two-Dimensional NMR Spectroscopy. – Multiple-Quantum NMR Spectroscopy. – Magnetic Shielding Tensor. – Spin-Lattice Relaxation. – Appendix. – References. – Subject Index.

C. Fest, K.-J. Schmidt

The Chemistry of Organophosphorus Pesticides

2nd revised edition. 1982. 44 figures. X, 360 pages
ISBN 3-540-11303-7

Contents: Introduction. – General Section. – Chemical Section. – Biochemistry. – Appendix.

Springer-Verlag
Berlin
Heidelberg
New York
Tokyo